OXFORD

UNIVERSITY PRESS

OXFORD PROBABILITY & STATISTICS

2

for Cambridge International AS & A Level

Series editor: David Rayner

James Nicholson

Oxford excellence for Cambridge AS & A Level

OXFORD
UNIVERSITY PRESS

Great Clarendon Street, Oxford, OX2 6DP, United Kingdom

Oxford University Press is a department of the University of Oxford.
It furthers the University's objective of excellence in research,
scholarship, and education by publishing worldwide. Oxford is a registered
trade mark of Oxford University Press in the UK and in certain other countries

British Library Cataloguing in Publication Data
Data available

978-0-19-830694-8

10 9 8 7 6 5 4 3 2

Paper used in the production of this book is a natural, recyclable product
made from wood grown in sustainable forests.
The manufacturing process conforms to the environmental regulations
of the country of origin.

Printed in Great Britain by CPI Group (UK) Ltd., Croydon CR0 4YY

The publisher would like to thank Cambridge International Examinations
for their kind permission to reproduce past paper questions.

*Cambridge International Examinations bears no responsibility for the example
answers to questions taken from its past papers which are contained in this
publication.*

*The questions, example answers, marks awarded and comments that appear in this
book were written by the authors. In examination, the way marks would be awarded
to answers like these may be different.*

Acknowledgements

The publisher would like to thank the following for permission to reproduce photographs:

p1: Norma Jean Gargasz/age footstock; **p2 (TL):** Rackermann/iStockphoto; p2 (TR) Maica/iStockphoto;
p2 (ML): dpa picture alliance/Alamy Stock Photo; **p2 (MR)** Tim Graham/Alamy Stock Photo; **p2 (BL):**
Echo/Gettyimages; **p2 (BR):** Richard Wear/age footstock; **p13 (T):** ERproductions Ltd/age footstock;
p13 (BL) Chungking/Shutterstock; **p13 (BL):** Phil Robinson/age footstock; p14 (T) LoloStock/
Shutterstock; **p14 (BL):** Dzimin/Fotolia; **p14 (BR):** Pixel Shepherd/age footstock; **p20:** Leigh Prather/
Shutterstock; **p24:** Gwright/Alamy Stock Photo; **p28:** Martin Plöb/age footstock; p47 (T) Bert de Ruiter/
Alamy Stock Photo; **p47 (M):** Vadim Petrakov/Shutterstock; **p47 (B):** Danita Delmont/Shutterstock;
p48: Caro Seeberg/Photoshot; p55 (TL) OJO_Images/iStockphoto; p55 (TR) Denis Kuvaev/Shutterstock;
p60: Barna Tanko/Shutterstock; **p62:** Lucky-photographer/Shutterstock; **p63:** Dariazu/Shutterstock;
p80: FloridaStock/Shutterstock; **p86 (BL):** James Steidl/Shutterstock; **p86(BR):** Photology1971/
Shutterstock; **p112 (T):** f11photo/Shutterstock; **p112 (B):** Ei Katsumata/age footstock; p113 Frank
Vetere/Alamy Stock Photo; **p114:** Ian Murray/age footstock; **p136:** Javier Larrea/age footstock;
p160: Sarymsakov Andrey/Shutterstock; **p176 (TL):** MR1805/iStockphoto; **p176 (TR):** ssuaphoto/
iStockphoto; **p177:** AshDesign/Shutterstock

Cover Illustration by Ian Norris, Oxford University Press

Contents

Introduction

About this book

This book has been written to cover the **Cambridge International AS and A Level Mathematics (9709)** course, and is fully aligned to the syllabus.

In addition to the main curriculum content, you will find:

- 'Maths in real-life', showing how principles learned in this course are used in the real world.
- Chapter openers, which outline how each topic in the Cambridge 9709 syllabus is used in real-life.

The book contains the following features:

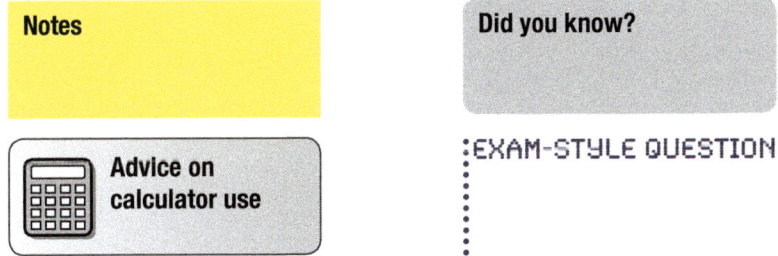

Throughout the book, you will encounter worked examples and a host of rigorous exercises. The examples show you the important techniques required to tackle questions. The exercises are carefully graded, starting from a basic level and going up to exam standard, allowing you plenty of opportunities to practise your skills. Together, the examples and exercises put maths in a real-world context, with a truly international focus.

At the start of each chapter, you will see a list of objectives that are covered in the chapter. These objectives are drawn from the Cambridge AS and A Level syllabus. Each chapter begins with a *Before you start* section and finishes with a *Summary exercise* and *Chapter summary*, ensuring that you fully understand each topic.

A *Review exercise* is placed after chapters 3, 6 and 9. These comprise a host of exam questions which cover the topics from the previous three chapters and are in no particular order of difficulty.

The answers given at the back of the book are concise. However, when answering exam-style questions, you should show as many steps in your working as possible. All exam-style questions have been written by the authors.

About the author

James Nicholson is an experienced teacher of mathematics at secondary level, having taught for 12 years at Harrow School as well as spending 13 years as Head of Mathematics in a large Belfast grammar school. He is the author of two A Level statistics texts, and editor of the *Concise Oxford Dictionary of Mathematics*. He has also contributed to a number of other sets of curriculum and assessment materials, is an experienced examiner and has acted as a consultant for UK government agencies on accreditation of new specifications.

James ran schools workshops for the Royal Statistical Society for many years, and has been a member of the Schools and Further Education Committee of the Institute of Mathematics and its Application since 2000, including 6 years as chair, and is currently a member of the Outer Circle group for the Advisory Committee on Mathematics Education. He has served as a Vice-President of the International Association for Statistics Education for 4 years, and is currently Chair of the Advisory Board to the International Statistical Literacy Project.

A note from the author

The aim of this book is to help students prepare for the Statistics 2 unit of the Cambridge International AS and A Level Mathematics syllabus, though it may also be found to be useful in providing support material for other AS and A Level courses. The book contains a large number of practice questions, many of which are exam-style, in addition to questions from past Cambridge examination papers.

In writing the book I have drawn on my experiences of teaching Mathematics, Statistics and Further Mathematics to A Level over many years as well as on my experience as an examiner, and discussion with statistics educators from many countries at international conferences.

1 The Poisson distribution

The Poisson distribution can be used to (at least approximately) model a large number of natural and social phenomena. You might not expect the number of photons arriving at a cosmic ray observatory, the number of claims made to an insurance company, the number of earthquakes of a given intensity and the number of atoms decaying in a radioactive material to have much in common, but they are all examples of this distribution. The photo is of VERITAS – Very Energetic Radiation Telescope Array in Arizona – which is helping to shape our understanding of how subatomic particles like photons are accelerated to extremely high energy levels.

Objectives

After studying this chapter you should be able to:

- calculate probabilities for the distribution $Po(\mu)$;
- use the fact that if $X \sim Po(\mu)$ then the mean and variance of X are each equal to μ;
- understand the relevance of the Poisson distribution to the distribution of random events, and use the Poisson distribution as a model;

Before you start

You should know how to:

1. Use your calculator to work out values of exponential functions.

   Find the value of $e^{-2.5}$

   $e^{-2.5} = 0.0821$

2. Substitute values into more complex formulae

   Find the value of $p = \dfrac{e^{-2.5} \times 2.5^4}{4!}$

   $\dfrac{e^{-2.5} \times 2.5^4}{4!} = \dfrac{0.0821 \times 39.06}{24} 0.133$

Skills check:

1. Find the value of:

   a) e^{-3}

   b) $e^{-2.1}$

2. Find the value of $p = \dfrac{e^{-3} \times 3^5}{5!}$

1.1 Introducing the Poisson distribution

Think about the following random variables:

- The number of dandelions in a square metre of a piece of open ground.
- The number of errors in a page of a typed manuscript.
- The number of cars passing a point on a motorway in a minute.
- The number of telephone calls received by a company switchboard in half an hour.
- The number of lightning strikes in an area over a year.

Do they have any features in common? Does any one of them stand out as being rather different?

The behaviour of five of these random variables follow the *Poisson distribution*.

Formally, the conditions are that
i)   events occur at random
ii) events occur independently of one another
iii) the average rate of occurrences remains constant
iv) there is zero probability of simultaneous occurrences.

The Poisson distribution is defined as

$$P(X = r) = \frac{e^{-\lambda}\lambda^r}{r!} \text{ for } r = 0, 1, 2, 3 \dots \dots$$

You need to have a value for λ in order for this to make sense, so there is a **family** of Poisson distributions but there is only one parameter, λ, which is the mean number of occurrences in the time period (or length, or area) being considered.

Like the Binomial, you can write the Poisson distribution as $X \sim \text{Po}(\lambda)$.

In the first photo of the middle pair the traffic is free flowing and cars can overtake when they want to but in the second photo there is much less randomness because the traffic is so heavy it is all travelling at almost the same speed.

You will look later in this chapter in more detail at cases where the four conditions listed here are not met exactly.

Note: The syllabus uses μ for the parameter whereas this book uses λ. Either are acceptable

---

Example 1

If $X \sim \text{Po}(3)$ find $P(X = 2)$.

$P(X = 2) = \frac{e^{-3}3^2}{2!} = 0.224 (3 \text{ s.f.})$

Example 2

The number of cars passing a point on a road during a five minute period may be modelled by the Poisson distribution with parameter 4.

Find the probability that in a five minute period
i)   2 cars go past   ii) fewer than 3 cars go past.

$X \sim \text{Po}(4);$

i) $P(X = 2) = \frac{e^{-4}4^2}{2!} = 0.146525\dots = 0.147 (3 \text{ s.f.})$

ii) $P(X = 0) = \frac{e^{-4}4^0}{0!} = 0.01831\dots = 0.0183 (3 \text{ s.f.})$

$P(X = 1) = \frac{e^{-4}4^1}{1!} = 0.07326\dots = 0.0733 (3 \text{ s.f.})$

$P(X < 3) = 0.01831 \dots + 0.07326\dots + 0.146525\dots$

$= 0.238 (3 \text{ s.f.})$

Mathematical note: It is not immediately obvious from the mathematics you cover in this course that the form of the Poisson constitutes a probability distribution – remember in S1 chapter 5 this requires all probabilities to be non-negative (which they obviously all are here) but also that the sum of the probabilities is 1.

$P(X = r) = \dfrac{e^{-\lambda}\lambda^r}{r!}$ for $r = 0, 1, 2, 3.... ....$ is a probability distribution

because $\displaystyle\sum_{r=0}^{\infty}\left(\dfrac{\lambda^r}{r!}\right) = 1 + \lambda + \dfrac{\lambda^2}{2!} + \dfrac{\lambda^3}{3!} + \dfrac{\lambda^4}{4!} + ... = e^{\lambda}$ – this is an example

of an advanced topic in Pure Maths where functions like exponentials, logarithms and the trigonometric functions have (infinite) power series forms. Truncated forms of these infinite series are how electronic calculators obtain values of these functions.

Exercise 1.1

1. If $X \sim Po(2)$ find   **i)** $P(X = 1)$ **ii)** $P(X = 2)$ **iii)** $P(X = 3)$

2. If $X \sim Po(1.8)$ find   **i)** $P(X = 0)$ **ii)** $P(X = 1)$ **iii)** $P(X = 2)$

3. If $X \sim Po(5.3)$ find   **i)** $P(X = 3)$ **ii)** $P(X = 5)$ **iii)** $P(X = 7)$

4. If $X \sim Po(0.4)$ find   **i)** $P(X = 0)$ **ii)** $P(X = 1)$ **iii)** $P(X = 2)$

5. If $X \sim Po(2.15)$ find   **i)** $P(X = 2)$ **ii)** $P(X = 4)$ **iii)** $P(X = 6)$

6. If $X \sim Po(3.2)$ find   **i)** $P(X = 2)$ **ii)** $P(X \le 2)$ **iii)** $P(X \ge 2)$

7. The number of telephone calls arriving at an office switchboard in a five minute period may be modelled by a Poisson distribution with parameter 3.2. Find the probability that in a five minute period

   a) exactly 2 calls are received;

   b) more than 2 calls are received.

8. The number of accidents which occur on a particular stretch of road in a day may be modelled by a Poisson distribution with parameter 1.3. Find the probability that on a particular day

   a) exactly 2 accidents occur on that stretch of road;

   b) fewer than 2 accidents occur.

1.2 The role of the parameter of the Poisson distribution

The mean number of events in an interval of time or space is proportional to the size of the interval.

Example 2 in Section 1.1 looked at the number of cars passing a point on a road during a five minute period. This may be modelled by the Poisson distribution with parameter 4.

In this case, the number of cars passing that point in a twenty minute period may be modelled by the Poisson distribution with parameter 16, and in a one minute period may be modelled by the Poisson distribution with parameter 0.8.

If the conditions for a Poisson distribution are satisfied in a given period, they are also satisfied for periods of different length.

If the average rate of occurrences remains constant, then the mean number of occurrences in an interval will be proportional to the length of the interval.

Example 3

The number of accidents in a week on a stretch of road is known to follow a Poisson distribution with parameter 2.1.

Find the probability that:

a) in a given week there is 1 accident

b) in a two week period there are 2 accidents

c) there is 1 accident in each of two successive weeks.

a) In 1 week, the number of accidents follows a Po(2.1) distribution,

so the probability of 1 accident $= \dfrac{e^{-2.1}2.1^1}{1!} = 0.257(3\,\text{s.f.})$

b) In 2 weeks, the number of accidents follows a Po(4.2) distribution,

so the probability of 2 accidents $= \dfrac{e^{-4.2}4.2^2}{2!} = 0.132(3\,\text{s.f.})$

c) This can not be done directly as a Poisson since it says what has to happen in each of two time periods, but these are the outcomes considered in part **a**.

So the probability this happens in two successive weeks is $\left(\dfrac{e^{-2.1}2.1^1}{1!}\right)^2 = 0.0661(3\,\text{s.f.})$

This is considerably less than the probability in part **b**, and this is because '1 accident in each of two successive weeks' will give '2 accidents in a two week period' but so will having none then two, or having two and then none, so the total probability that in a two week period there are 2 accidents must be higher than specifying there will be 1 in each week.

Example 4

The number of flaws in a metre length of dress material is known to follow a Poisson distribution with parameter 0.4.

Find the probabilities that

a) there are no flaws in a 1 metre length

b) there is 1 flaw in a 3 metre length

c) there is 1 flaw in a piece of material which is half a metre long.

a) $X \sim \text{Po}(0.4); \quad \Rightarrow P(X = 0) = \dfrac{e^{-0.4} \times 0.4^0}{0!} = 0.670 (3\,\text{s.f.})$

b) $Y \sim \text{Po}(1.2); \quad \Rightarrow P(Y = 1) = \dfrac{e^{-1.2} \times 1.2^1}{1!} = 0.361 (3\,\text{s.f.})$

c) $Z \sim \text{Po}(0.2); \quad \Rightarrow P(Z = 1) = \dfrac{e^{-0.2} \times 0.2^1}{1!} = 0.164 (3\,\text{s.f.})$

> It is good practice to define new variable names when the interval changes. While all three of these calculations relate to the same basic situation, they all use different Poisson distributions and this is a simple way to stop confusion arising.

Exercise 1.2

1. The number of telephone calls arriving at an office switchboard in a five minute period may be modelled by a Poisson distribution with parameter 1.4. Find the probability that in a ten minute period

   a) exactly 2 calls are received;

   b) more than 2 calls are received.

2. The number of accidents which occur on a particular stretch of road in a day may be modelled by a Poisson distribution with parameter 0.4. Find the probability that during a week (7 days)

   a) exactly 2 accidents occur on that stretch of road;

   b) fewer than 2 accidents occur.

3. The number of letters delivered to a house on a day may be modelled by a Poisson distribution with parameter 0.8.

   a) Find the probability that there are 2 letters delivered on a particular day.

   b) The home owner is away for 3 days. Find the probability that there will be more than 2 letters waiting for him when he gets back.

4. The number of errors on a page of a booklet can be modelled by a Poisson distribution with parameter 0.2.

   a) Find the probability that there is exactly 1 error on a given page.

   b) A section of the booklet has 7 pages. Find the probability that there are no more than 2 errors in the section.

   c) The booklet has 25 pages altogether. Find the probability that the booklet contains exactly 6 errors altogether.

5. The number of people calling a car breakdown service can be modelled by a Poisson distribution, and the service has an average of 6 calls per hour. Find the probability that in a half hour period
   a) exactly 2 calls are received;
   b) more than 2 calls are received.

1.3 The recurrence relation for the Poisson distribution

This is not directly in the course, but it is a useful property of the Poisson to be aware of, and gives some insight into why the Poisson distribution has the shape that it does.

You can calculate probabilities for a Poisson distribution in sequence using a recurrence relation.

Example 5

If $X \sim Po(\lambda)$

a) write down the probability that
   i) $X = 3$ and   ii) $X = 4$.
b) write $P(X = 4)$ in terms of $P(X = 3)$

a) i) $\dfrac{e^{-\lambda} \times \lambda^4}{4!}$ ii) $\dfrac{e^{-\lambda} \times \lambda^3}{3!}$

b) $\dfrac{e^{-\lambda} \times \lambda^4}{4!} = \left(\dfrac{e^{-\lambda} \times \lambda^3}{3!} \right) \times \dfrac{\lambda}{4}$

so $P(X = 4) = P(X = 3) \times \dfrac{\lambda}{4}$ $\qquad$ $4! = 4 \times 3 \times 2 \times 1 = 4 \times (3 \times 2 \times 1) = 4 \times 3!$

The general relationship is $P(X = k + 1) = \dfrac{\lambda}{k+1} \times P(X = k)$

These graphs show the probability distributions for different values of λ and what effect this has on the shape of a particular Poisson distribution.

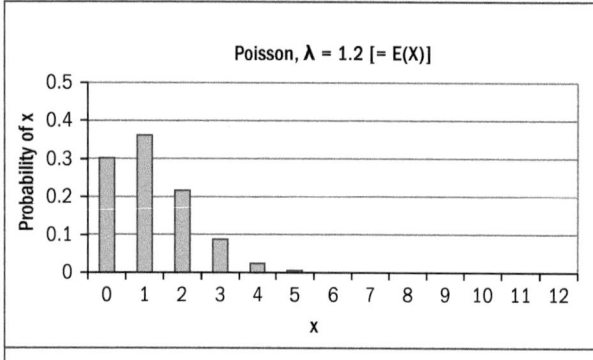

All Poisson variables have an outcome space which is all of the non-negative integers. However, when λ is relatively low, the probabilities tail off very quickly.

$$\frac{1.2}{1} = 1.2; \quad \frac{1.2}{2} = 0.6; \quad \frac{1.2}{3} = 0.4; \quad \frac{1.2}{4} = 0.3\ldots\ldots$$

so the initial probability that $X = 0$ is multiplied by 1.2, then 0.6, then 0.4, 0.3 …. and so the mode is when $X = 1$.

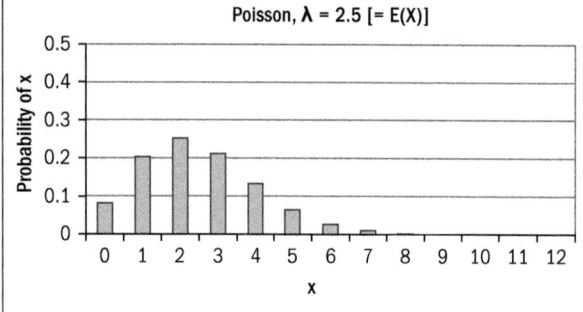

Here λ is larger than in the previous graph, and the peak has moved across to the right.
For values of X which are less than λ the probability is higher than the previous probability, but once x is $> \lambda$ the probabilities start to decrease.
More values of x have a noticeable probability, so the highest individual probability is not as large as it was in the previous graph and the distribution is more spread out.

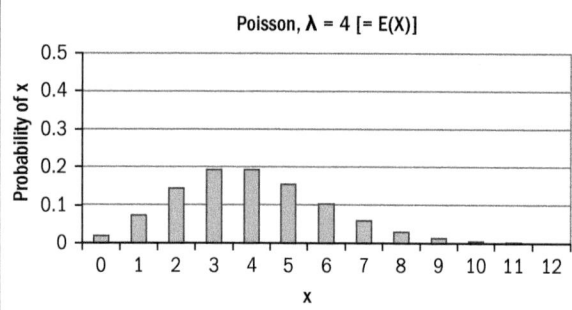

What happens when λ is an integer?
Here $P(X = 4) = P(X = 3) \times \frac{4}{4} = P(X = 3)$ and the distribution has two modes – at 3 and 4.
Generally, the mode of the Poisson (λ) distribution is at the integer below λ when λ is not an integer and there are 2 modes (at λ and $\lambda - 1$) when it is an integer.

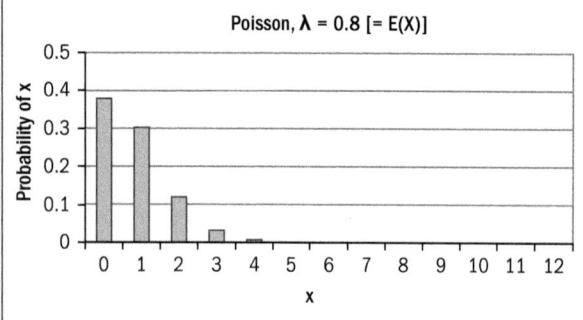

$\lambda < 1$ is a special case.
Here even the first time the recurrence relation is used you are multiplying by < 1, so the mode will be 0 and the probability distribution is strictly decreasing for all values of x.

The general forms for the probabilities of 0 and 1 for a Poisson distribution are

$$P(X = 0) = \frac{e^{-\lambda} \times \lambda^0}{0!} = e^{-\lambda} \text{ and } P(X = 1) = \frac{e^{-\lambda} \times \lambda^1}{1!} = \lambda e^{-\lambda}$$

Example 6

$X \sim Po(\lambda)$ and $P(X = 6) = 2 \times P(X = 5)$. Find the value of λ.

$P(X = 6) = \frac{\lambda}{6} \times P(X = 5)$ so $\frac{\lambda}{6} = 2$ and $\lambda = 12$.

Example 7

$X \sim Po(5.8)$. State the mode of X.

Since 5.8 is not an integer, the mode is the integer below it
i.e. the mode is 5.

Exercise 1.3

1. $X \sim Po(2.5)$.
   a) Write down an expression for $P(X = 4)$ in terms of $P(X = 3)$.
   b) If $P(X = 3) = 0.214$, calculate the value of your expression in part **a**.
   c) Calculate $P(X = 4)$ directly and check it is the same as your answer to **b**.
   d) What is the mode of X?

2. $X \sim Po(5)$
   a) Write down an expression for $P(X = 5)$ in terms of $P(X = 4)$
   b) Explain why X has two modes at 4 and 5.

3. $X \sim Po(\lambda)$ and $P(X = 4) = 1.2 \times P(X = 3)$.
   a) Find the value of λ.
   b) What is the mode of X?

1.4 Mean and variance of the Poisson distribution

> If $X \sim \text{Po}(\lambda)$, then $E(X) = \lambda$; $\text{Var}(X) = \lambda \Rightarrow$ st. dev $(\sigma) = \sqrt{\lambda}$.

A special property of the Poisson distribution is that the mean and variance are always equal.

Example 8

The number of calls arriving at a company's switchboard in a ten minute period can be modelled by a Poisson distribution with parameter 3.5.

Give the mean and variance of the number of calls which arrive in

i) ten minutes   ii) an hour   iii) five minutes

i)   here $\lambda = 3.5$ so the mean and variance will both be 3.5

ii) here $\lambda = 21$ ($= 3.5 \times 6$) so the mean and variance will both be 21

iii) here $\lambda = 1.75$ ($= 3.5 \div 2$) so the mean and variance will both be 1.75

Example 9

A dual carriageway has one lane blocked off because of roadworks.

The number of cars passing a point in a road in a number of one minute intervals is summarised in the table.

number of cars	0	1	2	3	4	5	6
frequency	3	4	4	25	30	3	1

a) Calculate the mean and variance of the number of cars passing in one minute intervals.

b) Is the Poisson likely to provide an adequate model for the distribution of the number of cars passing in one minute intervals?

a) $\sum f = 70$, $\sum xf = 228$ $\sum x^2 f = 836$, so $\bar{x} = \frac{228}{70} = 3.26 (3\,\text{s.f.})$

and $\text{Var}(X) = \dfrac{\sum x^2 f}{\sum f} - \bar{x}^2 = \dfrac{836}{70} - \left(\dfrac{228}{70}\right)^2 = 1.33 (3\,\text{s.f.})$

b) The mean and variance are not numerically close so it is unlikely the Poisson will be an adequate model (with only one lane open for traffic, overtaking cannot happen on this stretch of the road and the numbers of cars will be much more consistent than would happen in normal circumstances – hence the variance is much lower than would be expected if the Poisson model did apply).

Derivation of Mean and Variance of the Poisson distribution*

You must be able to use these results but are not required to be able to prove them – they are included here for completeness, and as a nice manipulation using the power series expression for the exponential function.

$$X \sim \text{Po}(\lambda) \Leftrightarrow \Pr\{X = k\} = \frac{e^{-\lambda}\lambda^k}{k!}$$

$$E(X) = \sum_{k=0}^{\infty} k \times \frac{e^{-\lambda}\lambda^k}{k!} = 0 \times \frac{e^{-\lambda}\lambda^0}{0!} + \sum_{k=1}^{\infty} k \times \frac{e^{-\lambda}\lambda^k}{k!}$$

$$= \lambda \times \sum_{k=1}^{\infty} \frac{e^{-\lambda}\lambda^{k-1}}{(k-1)!} \quad - \text{ cancelling } k, \text{ after discarding the zero case.}$$

$$= \lambda$$

$$E(X^2) = \sum_{k=0}^{\infty} k^2 \times \frac{e^{-\lambda}\lambda^k}{k!} = \lambda \times \sum_{k=1}^{\infty} k \times \frac{e^{-\lambda}\lambda^{k-1}}{(k-1)!} = \lambda \times \sum_{k=1}^{\infty} (k-1+1) \times \frac{e^{-\lambda}\lambda^{k-1}}{(k-1)!}$$

$$= \lambda^2 \times \sum_{k=2}^{\infty} \frac{e^{-\lambda}\lambda^{k-2}}{(k-2)!} + \lambda \times \sum_{k=1}^{\infty} \frac{e^{-\lambda}\lambda^{k-1}}{(k-1)!}$$

$$= \lambda^2 + \lambda$$

Then $\text{Var}(X) = \lambda^2 + \lambda - \lambda^2 = \lambda$

Exercise 1.4

1. If $X \sim \text{Po}(3.2)$ find **i)** $E(X)$ **ii)** $\text{Var}(X)$

2. If $X \sim \text{Po}(49)$ find the mean and standard deviation of X.

3. $X \sim \text{Po}(3.6)$

   a) Find the mean and standard deviation of X.

   b) Find $P(X > \mu)$, where $\mu = E(X)$.

   c) Find $P(X > \mu + 2\sigma)$, where σ is the standard deviation of X.

   d) Find $P(X < \mu - 2\sigma)$.

4. X is the number of telephone calls arriving at an office switchboard in a ten minute period. X may be modelled by a Poisson distribution with parameter 6.

   a) Find the mean and standard deviation of X.

   b) Find $P(X > \mu)$, where $\mu = E(X)$.

   c) Find $P(X > \mu + 2\sigma)$, where σ is the standard deviation of X.

   d) Find $P(X < \mu - 2\sigma)$.

5. Compare your answers to part **d)** of Q3 and Q4.

1.5 Modelling with the Poisson distribution

The Poisson describes the number of occurrences in a fixed period of time or space if the events occur independently of one another, at random and at a constant average rate.

Standard examples of Poisson processes in real life include:
radioactive emissions, traffic passing a fixed point, telephone calls or letters arriving, and accidents occurring.

Example 10

The maternity ward of a hospital wanted to work out how many births would be likely to happen during a night.

The hospital has 3000 deliveries each year, so if these happen randomly around the clock 1000 deliveries would occur between the hours of midnight and 8.00 a.m. This is the time when many staff are off duty and it is important to ensure that there will be enough people to cope with the workload on any particular night.

The average number of deliveries per night is $\frac{1000}{365}$, which is 2.74.

From this average rate the probability of delivering 0, 1, 2, etc. babies each night can be calculated using the Poisson distribution. If X is a random variable representing the number of deliveries per night, Some probabilities are:

$$P(X = 0) = 2.74^0 \times \frac{e^{-2.74}}{0!} = 0.065$$

$$P(X = 1) = 2.74^1 \times \frac{e^{-2.74}}{1!} = 0.177$$

$$P(X = 2) = 2.74^2 \times \frac{e^{-2.74}}{2!} = 0.242$$

$$P(X = 3) = 2.74^3 \times \frac{e^{-2.74}}{3!} = 0.221$$

i)   On how many days in the year would 5 or more deliveries be likely to occur?

ii) Over the course of one year, what is the greatest number of deliveries likely to occur at least once?

iii) Why might the pattern of deliveries *not* follow a Poisson distribution?

· ·

i) $52 = 365 \times P(X \geq 5)$

ii) 8 – the largest value for which the probability is greater than $\frac{1}{365}$

iii) If deliveries were not random throughout the 24 hours.

   e.g. If a lot of women had labour induced or had elective caesareans done during the day.

Did you know?

An elective caesarean is planned in advance for some births which are expected to be difficult.

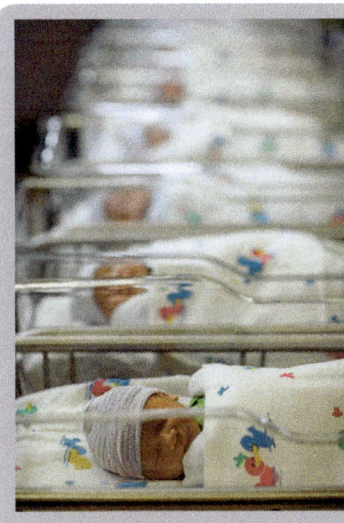

In this real life example, deliveries in fact followed the Poisson distribution very closely, and the hospital was able to predict the workload accurately.

The conditions for the Poisson distribution are that
i) events occur at random
ii) events occur independently of one another
iii) the average rate of occurrences remains constant
iv) there is zero probability of simultaneous occurrences.

Be careful:
Some change in the underlying conditions may alter the nature of the distribution, e.g. Traffic observed close to a junction, or where there are lane restrictions and traffic is funnelled into a queue travelling at constant speed.

The underlying conditions may be distorted by interference from other effects,

e.g. if a birthday or Christmas occurs during the period considered then the Poisson conditions would not be reasonable for the arrival of letters for instance.

Randomness or independence may be lost due to a difference in the average rate of occurrences,

e.g. the rate of traffic accidents occurring would be expected to vary somewhat as road conditions vary.

Example 11

The number of cyclists passing a remote village post-office during the day can be modelled as a Poisson random variable. On average two cyclists pass by in an hour.

a) What is the probability that
  i) no cyclist passes
  ii) more than 3 cyclists pass by between 10.00 and 11.00 am?

b) What is the probability that exactly one passes by while the shop-keeper is on a 20 minute tea-break?

c) What is the probability that more than 3 cyclists pass by in an hour exactly once in a six hour period?

a) In an hour (parts **i** and **ii**) $\lambda = 2$.

   i) $P(X = 0) = 0.1353$

   ii) $P(X > 3) = 1 - P(X \leq 3) = 1 - 0.8571 = 0.1429$

b) In a 20 minute period $= (\frac{1}{3}$ of an hour), the mean number of cyclists will be $2 \times \frac{1}{3} = \frac{2}{3}$

$$P(\text{exactly one}) = \frac{e^{-\frac{2}{3}}\left(\frac{2}{3}\right)^1}{1!} = 0.342 \ (3 \text{ s.f.}).$$

c) The situation is that of a Binomial distribution – there are 6 'trials', the number of cyclists in each hour is independent of the other periods, and the probability of more than 3 in an hour remains the same for all the 6 hour long periods, i.e. if Y = number of times that more than 3 cyclists pass by in an hour exactly once in a six hour period $Y \sim B(6, 0.1429)$ (using the probability calculated in part a **ii**).

> This cannot be treated as a single Poisson with parameter 12 since it specifies a particular event to be considered in each one hour time period separately.

$P(Y = 1) = 6 \times 0.1429^1 \times (1 - 0.1429)^5 = 0.397 \ (3 \text{ s.f.}).$

Example 12

At a certain harbour the number of boats arriving in a fifteen minute period can be modelled by a Poisson distribution with parameter 1.5

a) Find the probability that exactly six boats will arrive in a period of an hour.

b) Given that exactly six boats arrive in a period of an hour, find the conditional probability that there are twice as many arrive in the second half hour as arrive in the first half hour.

a) in an hour the average number of boats arriving is 6, so

$$P(6 \text{ boats arrive in an hour}) = \frac{e^{-6} 6^6}{6!} = 0.161$$

b) if twice as many arrive in the second half hour, then there needs to be 2 in a half hour period and then 4 in the next half hour, so

P(2 boats arrive in half hour, then 4 boats in next half hour)

$$= \frac{e^{-3} 3^2}{2!} \times \frac{e^{-3} 3^4}{4!} = 0.224 \times 0.168 = 0.0376$$

Then the conditional probability is

$$P(2 \text{ then } 4 \text{ in half hour} \mid 6 \text{ boats arrive in an hour}) = \frac{0.0376}{0.161} = 0.234$$

Exercise 1.5

1. For the following random variables state whether they can be modelled by a Poisson distribution.

   If they can, give the value of the parameter λ, if they cannot then explain why.

   a) The average number of cars per minute passing a point on a road is 12.

   The traffic is flowing freely.

   X = number of cars which pass in a 15 second period.

   b) The average number of cars per minute passing a point on a road is 14.

   There are roadworks blocking one lane of the road.

   X = number of cars which pass in a 30 second period.

   c) Amelie normally gets letters at an average rate of 1.5 per day.

   X = number of letters Amelie gets on December 22nd.

   > Amelie lives in a country where Christmas is a major festival on December 25 each year.

   d) A petrol station which stays open all the time gets an average of 832 customers in a 24 hour time period.

   X = number of customers in a quarter of an hour at the petrol station.

   e) An A&E department in a hospital treats 32 patients an hour on average.

   X = number of patients treated between 5 pm and 7 pm on a Friday evening.

2. For the following situations state what assumptions are needed if a Poisson distribution is to be used to model them, and give the value of λ that would be used.

   You are **not** expected to do any calculations!

   a) On average defects in a roll of cloth occur at a rate of 0.2 per metre.

   How many defects are there in a roll which is 8 m long?

   b) On average defects in a roll of cloth occur once in 2 metres.

   How many defects are there in a roll which is 8 m long?

   c) A small shop averages 8 customers per hour.

   How many customers does it have in twenty minutes?

3. An explorer thinks that the number of mosquito bites he gets when he is in the jungle will follow a Poisson distribution.

   The explorer records the number of mosquito bites he gets in the jungle during a number of hour long periods, and the results are summarised in the table

number of bites	0	1	2	3	4	5	6	≥ 7
frequency	3	7	9	6	6	3	1	0

a) Calculate the mean and variance of the number of bites the explorer gets in an hour in the jungle.

b) Do you think the Poisson is a good model for the number of bites the explorer gets in an hour in the jungle?

4. The number of emails Serena gets can be modelled by a Poisson distribution with a mean rate of 1.5 per hour.

   a) **i)** What is the probability that Serena gets no emails between 4 pm and 5 pm?

       ii) What is the probability that Serena gets more than 2 emails between 4 pm and 5 pm?

       iii) What is the probability that Serena gets one email between 6 pm and 6.20 pm?

   b) What is the probability that Serena gets more than 2 emails in an hour exactly twice in a five hour period?

   c) Would it be sensible to use the Poisson distribution to find the probability that Serena gets no emails between **4 am** and **5 am**?

5. The number of lightning strikes in the neighbourhood of a campsite in a week can be modelled by a Poisson distribution with parameter 1.5

   a) Find the probability that there is exactly one lightning strike in the neighbourhood in a given week.

   b) Alejandra spends three weeks at the campsite. Find the probability that there are exactly three lightning strikes in the neighbourhood during her holiday.

   c) Given that the neighbourhood has exactly three lightning strikes during her holiday, find the conditional probability that each week has exactly one strike.

Summary exercise 1

1. If $X \sim P(1.75)$ find

   a) $P(X = 2)$

   b) $P(X \leq 2)$

   c) $P(X \geq 2)$

2. If $X \sim Po(3.2)$

   a) find **i)** $P(X = 0)$ **ii)** $P(X = 1)$

       iii) $P(X > 2)$

   b) For X, write down the **i)** mean
       ii) variance   **iii)** standard deviation

   c) Explain why the mode of X is 3.

d) Find

   i) $P(X < \mu)$ **ii)** $P(|X - \mu| < \sigma)$
   where μ is the mean and σ is the standard deviation of X.

3. If $X \sim Po(6.4)$

   a) find **i)** $P(X \leq 3)$ **ii)** $P(X = 6)$

   b) For X, write down the **i)** mean
       ii) variance   **iii)** standard deviation

   c) Write down the mode of X.

   d) Find **i)** $P(X < \mu - \sigma)$

       ii) $P(|X - \mu| < \sigma)$ where μ is the mean and σ is the standard deviation of X.

4. The number of telephone calls arriving at Sharma's home in a fifteen minute period may be modelled by a Poisson distribution with parameter 0.4. Find the probability that in an hour
   a) exactly 2 calls are received
   b) more than 2 calls are received.

5. The number of accidents which occur on a particular stretch of road in a day may be modelled by a Poisson distribution with parameter 1.6. Find the probability that on a particular day
   a) exactly 2 accidents occur on that stretch of road.
   b) less than 2 accidents occur.

6. X is the number of telephone calls arriving at an office switchboard in a ten minute period. X may be modelled by a Poisson distribution with parameter 4.
   a) Find the mean and standard deviation of X.
   b) Find $P(X > \mu)$, where $\mu = E(X)$.
   c) Find $P(|X - \mu| > \sigma)$, where σ is the standard deviation of X.

7. An urban safety officer thinks that the number of traffic accidents in an area will follow a Poisson distribution.

   The officer records the number of accidents in the area each week over a period of several months, and the results are summarised in the table.

number of accidents	0	1	2	3	4	5	6	≥7
frequency	5	1	1	3	5	2	1	0

   a) Calculate the mean and variance of the number of accidents in the area in a week,

   b) Do you think the Poisson is a good model for the number of accidents in the area in a week?

8. The number of errors on a page of a book can be modelled by a Poisson distribution with parameter 0.15.
   a) Find the probability that there is exactly 1 error on a given page.
   b) A chapter of the book has 20 pages. Find the probability that there are no more than 2 errors in the chapter.
   c) What is the most likely number of errors in the chapter?

EXAM-STYLE QUESTIONS

9. The number of errors on a page of the first proofs of a book can be modelled by a Poisson distribution with parameter 0.6.
   a) Find the probability that a page has exactly one error on it.
   b) Find the probability that a double page spread has exactly two errors on it.
   c) Given that a double page spread has exactly two errors on it, find the conditional probability that each page has exactly one error on it.

10. A shop sells spades. The demand for spades follows a Poisson distribution with mean 2.7 per week.
    a) Find the probability that the demand is exactly 2 spades in any one week.
    b) The shop has 4 spades in stock at the beginning of a week. Find the probability that this will be enough to satisfy the demand for spades in that week.

c) Given instead that there are n spades in stick, find, by trial and error, the least value of n for which the probability of not being able to satisfy the demand for spades in that week is less than 0.1.

11. The random variable X has the distribution Po(2.5). The random variable Y is defined by $Y = 2x$.

   a) Find the mean and variance of Y.

   b) Give a reason why the variable Y does not have a Poisson distribution.

12. People arrive randomly and independently at a supermarket checkout at an average rate of 2 people every 3 minutes.

   a) Find the probability that exactly 4 people arrive in a 5-minute period. [2]

   At another checkout in the same supermarket, people arrive randomly and independently at an average rate of 1 person each minute.

   b) Find the probability that a total of fewer than 3 people arrive at the two checkouts in a 3-minute period. [3]

Cambridge International AS and A Level Mathematics 9709 Paper 71 Q2 November 2010

13. The number of radioactive particles emitted per 150-minute period by some material has a Poisson distribution with mean 0.7.

   a) Find the probability that at most 2 particles will be emitted during a randomly chosen 10-hour period. [3]

   b) Find, in minutes, the longest time period for which the probability that no particles are emitted is at least 0.99. [5]

Cambridge International AS and A Level Mathematics 9709 Paper 71 Q2 November 2013

Chapter summary

- The Poisson distribution is defined as
$$P(X = r) = \frac{e^{-\lambda}\lambda^r}{r!} \text{ for } r = 0, 1, 2, 3\ldots\ldots$$

- The Poisson distribution has a single parameter, λ

- The Poisson distribution is often written as $X \sim P(\lambda)$.

- If $X \sim P(\lambda)$, then $E(X) = \lambda$; $Var(X) = \sigma^2 = \lambda \Rightarrow$ st. dev$(\sigma) = \sqrt{\lambda}$

- The conditions for the Poisson are:

   i) events occur at random

   ii) events occur independently of one another

   iii) the average rate of occurrences remains constant

   iv) there is zero probability of simultaneous occurrences.

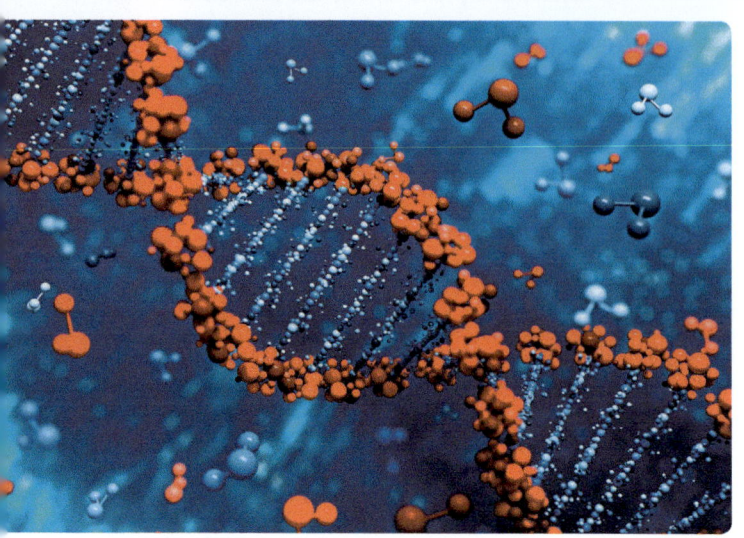

The Poisson provides a good approximation to large Binomial distributions under certain conditions.

For example, the number of genetic mutations in a stretch of DNA can be modelled well by the Poisson distribution – there is a lot of work currently being done to understand the processes involved in genetic mutations in both the plant and animal domains, with the possibility of significant medical advances in the treatment of diseases like cancer and Parkinson's.

Objectives

After studying this chapter you should be able to:

- use the Poisson distribution as an approximation to the binomial distribution where appropriate ($n > 50$ and $np < 5$, approximately);
- use the normal distribution, with continuity correction, as an approximation to the Poisson distribution where appropriate ($\mu > 15$, approximately).

Before you start

You should know how to:

1. Calculate probabilities using the Binomial distribution.

   $X \sim B(10, 0.3)$ Find $P(X = 2)$

   $$P(X = 2) = \binom{10}{2} 0.3^2 \times 0.7^8 = 0.233$$

2. Calculate probabilities using the Normal distribution

   $X \sim N(40, 15)$ Find $P(X < 44.2)$

   $$P(X < 44.2) = P\left(Z < \frac{44.2 - 40}{\sqrt{15}} = 1.084 \right)$$

   $$= 0.861$$

Skills check:

1. $X \sim B(40, 0.03)$ Find $P(X \le 2)$

2. $X \sim N(20, 20)$ Find $P(X < 17.1)$

2.1 Poisson as an approximation to the Binomial

In the last chapter of S1 you met the use of the Normal distribution as an approximation to the Binomial distribution, provided certain conditions were satisfied by the parameters n and p. Here we meet a second approximation to the Binomial.

> If $X \sim B(n, p)$ with n large ($n > 50$) and p close to 0 ($np < 5$) then $X \sim$ *approximately* $P(\lambda)$ with $\lambda = np$.

Here are some examples where the Binomial and Poisson distributions have the same mean:

If p is close to 0, then $q = 1 - p$ will be close to 1 and npq will be close to np.

np and npq are the mean and variance of the Binomial, and the Poisson has the property that the mean and variance are equal.

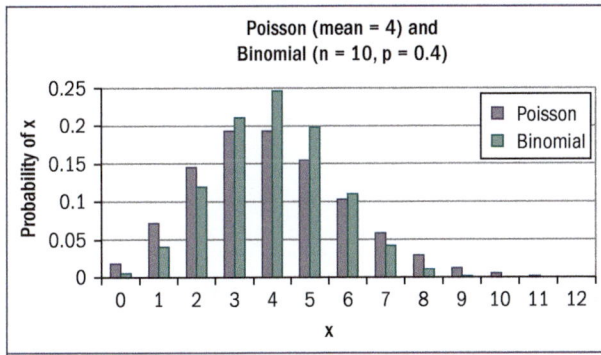

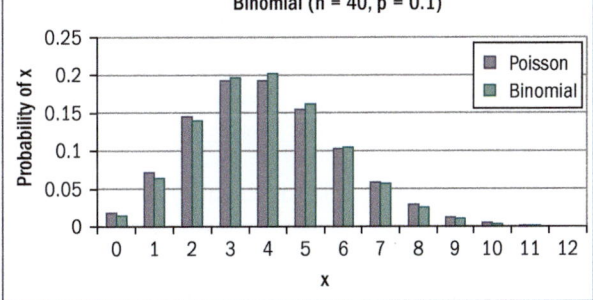

The mean of the binomial is 4 and the variance is 2.4. The two sets of probabilities are not particularly similar.	The variance of the binomial is now 3.6 (remember that the variance of the Poisson is 4). The agreement between the two sets of probabilities is now pretty strong.

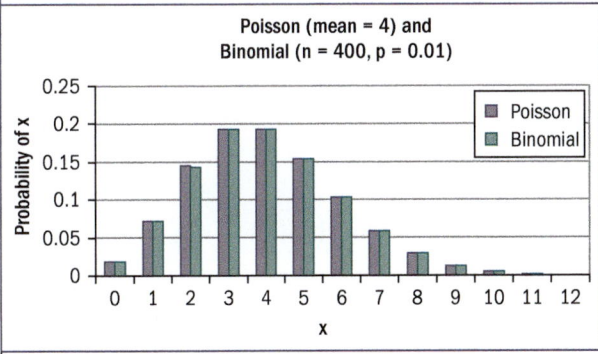

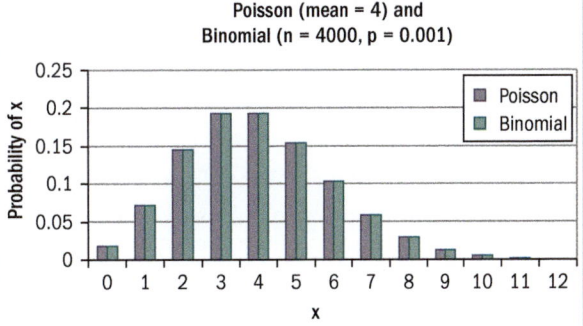

These two graphs both seem to show the Binomial and Poisson to be exactly the same – but they are not: while you cannot see any difference on this scale graphically, there are differences between the Binomial and the Poisson in both cases and the differences in the last case are much smaller than the differences when $n = 400$ and $p = 0.01$.

There is a fundamental difference in that the Poisson outcome space has no upper limit whereas the Binomial is bounded by the value of n. However, when n is large and p is small, the probabilities of high values of x are very small so this is not a problem (in the same way that the Normal can never provide an exact model for any physical measurements like heights or weights because the distribution cannot take negative values)

The use of the Poisson as an approximation to the Binomial improves as n increases and as p gets smaller.

Example 1

The probability that a component coming off a production line is faulty is 0.01.

a) If a sample of size 5 is taken, find the probability that exactly one of the components is faulty.

b) What is the probability that a batch of 250 of these components has more than 3 faulty components in it?

a) If X = number of faulty components in sample then $X \sim B(5, 0.01)$ and
$P(X = 1) = 5 \times 0.01 \times 0.99^4 = 0.0480 (3\,\text{s.f.})$

b) If Y = number of faulty components in the batch then $X \sim B(250, 0.01) \overset{approx}{\sim} P(2.5)$
and $P(Y > 3) = 1 - P(Y < 3) = 1 - 0.758 = 0.242$ (3 s.f.)

If you are working in a situation where p is close to 1, you can choose to count failures instead of successes and still construct an appropriate Poisson approximation.

Exercise 2.1

1. The proportion of defective pipes coming off a production line is 0.05.
   A sample of 60 pipes is examined.

   a) Using the exact Binomial distribution calculate the probabilities that there are

   i) 0   **ii)** 1   **iii)** 2

   iv) more than 2 defectives in the sample.

   b) Using an appropriate approximate distribution calculate the probabilities that there are

   i) 0   **ii)** 1   **iii)** 2

   iv) more than 2 defectives in the sample.

2. a) State the conditions under which a Poisson distribution may be used to approximate a Binomial distribution.

   b) 5% of the times a faulty ATM asks for a personal identification number (PIN number) it does not register the number entered correctly. If I enter my PIN correctly each time, what is the probability that the ATM will not register it correctly in 3 attempts?

c) Over a period of time, 90 attempts are made to enter a PIN. If all of the customers enter their PIN correctly, what is the probability that fewer than 3 of the attempts are not registered correctly.

3. In a small town, the football team claim that 95% of the people in town support them. If the claim is correct and a survey of 80 randomly chosen people asks whether they support the football team, find the probability that more than 75 people say they do.

4. A rare but harmless medical condition affects 1 in 200 people.

   a) In a cinema showing where 130 people attend, what is the probability that exactly one person has the condition?

   b) At a concert where the audience is 600, use an appropriate approximate distribution to find the probability that there are fewer than 5 people with the condition.

5. The Nutty Fruitcase party claim that 1 in 250 people support their policy to distribute free fruit and nut chocolate bars to children taking examinations.

   a) In an opinion poll which asks 1000 voters about a range of policies put forward by different parties what is the probability that

      i) no-one will support the Nutty Fruitcase party policy.

      ii) at least 5 people will support the policy.

   b) If the opinion poll had 7 people supporting the policy, does this mean that the Nutty Fruitcase party have underestimated the support there is for this policy?

2.2 The Normal approximation to the Poisson distribution

For large λ (> 15, approximately) you would often use a Normal approximation particularly when the probability of an interval is required E.g. $\Pr\{X \geq 15\}$ or $\Pr\{6 < X < 14\}$ since this is a single calculation in a continuous distribution but requires multiple calculations in a discrete distribution.

Remember that the Normal uses the **standard deviation** to calculate the z-score i.e. $z = \frac{x - \lambda}{\sqrt{\lambda}}$.

You must also include the continuity correction (which you met in S1 when using the Normal to approximate another discrete distribution – the Binomial).

The parameters used are the mean and variance of the Poisson i.e. $\mu = \sigma^2 = \lambda$.

Example 2

If $X \sim P(16)$ calculate $P(11 \leq X \leq 15)$

a) using the exact Poisson probabilities   **b)** by using a Normal approximation.

a) $P(11 \leq X \leq 15) = P(X = 11, 12, 13, 14, 15)$

$$= e^{-16}\left(\frac{16^{11}}{11!} + \frac{16^{12}}{12!} + \frac{16^{13}}{13!} + \frac{16^{14}}{14!} + \frac{16^{15}}{15!}\right) = 0.389$$

b) For the Poisson, $\mu = \lambda = 16$; $\sigma^2 = \lambda = 16$,

so use the N(16, 16) distribution to approximate the

P(10) distribution.

The continuity correction says $P(11 \leq X \leq 15) \overset{approx}{\approx} P(10.5 < Y < 15.5)$

where Y is the approximating Normal.

$$P(10.5 < Y < 15.5) = P\left(\frac{10.5 - 16}{\sqrt{16}} < Z < \frac{15.5 - 16}{\sqrt{16}}\right)$$

$$= P(-1.375 < Z < -0.125)$$

$$= \Phi(1.375) - \Phi(0.125) = 0.9155 - 0.5498 = 0.366$$

Example 3

The demand for a particular spare part in a car accessory
shop may be modelled by a Poisson distribution.
On average the demand per week for that part is 2.5.

a) The shop has 4 in stock at the start of one week.
What is the probability that they will not be able
to supply everyone who asks for that part during
the week.

b) The manager is going to be away for 6 weeks, and
wants to leave sufficient stock that there is no more than a 5% probability of running
out of any parts while he is away. How many of this particular spare part should
he have in stock when he leaves?

a) For the demand in a week, use the P(2.5) distribution, Then if the
demand is ≤ 4 the shop can supply all the customers.

$$P(X < 4) = 0.0821 + 0.2052 + 0.2565 + 0.2138 + 0.1336 = 0.8912$$

The probability of not being able to supply all the demand is $1 - 0.891 = 0.109$ (3 s.f.).

b) For the demand in 6 weeks, use the P(15) distribution,

which can be approximated by the N(15,15) distribution.

You need to find k so that $P(\text{demand is} \leq k))$ is > 0.95.

$\Phi^{-1}(0.95) = 1.6449$ so you need to find the smallest integer k which satisfies

$\dfrac{k-15}{\sqrt{15}} > 1.6449$, which is 22 (solution is $k > 21.4$).

Exercise 2.2

1. Let $X \sim P(\lambda)$ and $Y \sim N(\lambda, \lambda)$ where λ satisfies the conditions needed for Y to be used as an approximation for X.

   Write down the probability you need to calculate for Y (including the continuity correction) as the approximation for each of the following probabilities for X.

   a) $P(X < 16)$ **b)** $P(X > 22)$ **c)** $P(X \le 17)$ **d)** $P(45 \le X < 62)$.

2. Which of the following could reasonably be approximated by a Normal distribution? (for those which can, give the Normal which would be used).

   a) $X \sim Po(16)$ **b)** $X \sim Po(12.32)$ **c)** $X \sim Po(8.5)$

3. Use Normal approximations to calculate

   a) $P(X < 42)$ if $X \sim Po(49)$ **b)** $P(X \ge 9)$ if $X \sim Po(17.5)$ **c)** $P(X \ge 13)$ if $X \sim Po(18.4)$

   d) $P(25 < X \le 37)$ if $X \sim Po(21)$ **e)** $P(43 \le X \le 55)$ if $X \sim Po(51.34)$

4. For $X \sim Po(20)$

   a) calculate $P(17 < X < 23)$

        i) directly   **ii)** using a Normal approximation.

   b) i) What error is there in using the Normal approximation?

        ii) Express this as a percentage of the exact probability.

5. Explain briefly the conditions under which the Normal distribution can be used as an approximation to a Poisson distribution.

   If the conditions are satisfied, state what Normal distribution would be used to approximate the $X \sim Po(\lambda)$ distribution.

6. The number of accidents which occur on a particular stretch of road in a day may be modelled by a Poisson distribution with parameter 0.5.

   a) Find the probability that during a week (7 days) exactly 2 accidents occur on that stretch of road.

   b) Find the probability that during a month (30 days) at least 20 accidents occur on that stretch of road.

7. The number of letters delivered to a house on a day may be modelled by a Poisson distribution with parameter 1.5.

   a) Find the probability that there are 2 letters delivered on a particular day.

   b) The home owner is away for 3 weeks (18 days of post). Find the probability that there will be more than 25 letters waiting for him when he gets back.

8. The number of errors on a page of a book can be modelled by a Poisson distribution with parameter 0.15. The book has 167 pages. Find the probability that there are no more than 20 errors in the book.

Summary exercise 2

1. The proportion of defective pipes coming off a production line is 0.025. A random sample of 60 pipes are examined

   a) Using the exact Binomial distribution calculate the probabilities that there are

      i) 0   ii) 1   iii) 2

      iv) more than 2 defectives in the sample.

   b) Using an appropriate approximate distribution calculate the probabilities that there are

      i) 0   ii) 1   iii) 2

      iv) more than 2 defectives in the sample.

2. A rare medical condition affects 1 in 150 sheep.

   a) In a small farm holding with a flock of 180 sheep, what is the probability that exactly one sheep has the condition?

   b) A large farm has a flock of 500 sheep, use an appropriate approximate distribution to find the probability that there are less than 5 sheep with the condition.

3. A rare disease affects 1 in 2000 people on average.

   a) Use a suitable approximation to find the probability that, of a random sample of 7500 people in a city, more than 3 people have the disease.

   b) In a random sample of n people, the probability that no one has the disease is less than 0.01. Find the least possible value of n.

4. Customers arrive at the exchange and refunds desk in a store at a constant average rate of 1 every 2 minutes.

   a) State one condition for the number of customers arriving in a given period to be modelled by a Poisson distribution. Assume now that a Poisson distribution is a suitable model.

   b) Find the probability that exactly 4 customers will arrive during a randomly chosen 10 minute period.

   c) Find the probability that less than 3 customers will arrive during a randomly chosen 5 minute period.

   d) Find an estimate that fewer than 25 customers will arrive in a randomly chosen hour long period.

EXAM-STYLE QUESTIONS

5. A random variable X has the distribution $Po(2.5)$

   a) A random value of X is found.

      i) Find $P(X \geq 2)$

      ii) Find the probability that $X = 2$ given that $X \geq 2$

   b) Random samples of 150 values of X are taken.

      i) Describe fully the distribution of the sample mean

      ii) Find the probability that the mean of a random sample of size 150 is less than 2.4.

6. On average 3 people in every 10000 in Canada have a particular gene. A random sample of 4000 people in Canada is chosen. The random variable X denoted the number of people in the sample who have the gene. Use an approximating distribution to calculate the probability that there will be more than 2 people in the sample who have the gene.

7. 2% of biscuits on a production line are broken. Broken biscuits occur randomly. 180 biscuits are checked to see whether they are broken. Use a suitable approximation to find the probability that fewer than 4 are broken. [3]

Cambridge International A Level and AS Mathematics 9709 Paper 71 Q1 November 2009

8. A book contains 40 000 words. For each word, the probability that it is printed wrongly is 0.0001 and these errors occur independently. The number of words printed wrongly in the book is represented by the random variable X.

   a) State the exact distribution of X, including the values of any parameters. [1]

   b) State an approximate distribution for X, including the values of any parameters, and explain why this approximate distribution is appropriate. [3]

   c) Use this approximate distribution to find the probability that there are more than 3 words printed wronly in the book. [3]

Cambridge International A Level and AS Mathematics 9709 Paper 71 Q3 November 2010

9. The probability that a new car of a certain type has faulty brakes is 0.008. A random sample of 520 new cars of this type is chosen, and the number, X, having faulty brakes is noted.

   a) Describe fully the distribution of X and describe also a suitable approximating distribution. Justify this approximating distribution. [4]

   b) Use your approximating distribution to find

   i) $P(X > 3)$, [2]

   ii) the smallest value of n such that $P(X = n) > P(X = n + 1)$. [3]

Cambridge International A Level and AS Mathematics 9709 Paper 71 Q5 June 2013

10. Each computer made in a factory contains 1000 components. On average, 1 in 30 000 of these components is defective. Use a suitable approximate distribution to find the probability that a randomly chosen computer contains at least 1 faulty component. [4]

Cambridge International A Level and AS Mathematics 9709 Paper 71 Q1 November 2013

Chapter summary

- If $X \sim B(n, p)$ with n large ($n > 50$) and p close to 0 ($np < 5$) then X ~ *approximately* $Po(\lambda)$ with $\lambda = np$
- If $X \sim Po(\lambda)$ with $\lambda > 15$ (approximately) then X ~ *approximately* $N(\lambda, \lambda)$
- When the Poisson is approximated by a Normal distribution, a continuity correction must be used.

3 Linear combination of random variables

The real world is not simple; many things are made up of more than one component. It is often easier to model each component of a process separately than it is to try to produce a complex model of the whole process. Simulations then allow you to get a good idea of what the behaviour of the overall process would be. For example, simulating the number of passengers on a flight, and then the baggage and person weights would be easier to do separately.

Objectives

After studying this chapter you should be able to:

- use, in the course of solving problems, the results that:
  - $E(aX + b) = aE(X) + b$ and $\mathrm{Var}(aX + b) = a^2\mathrm{Var}(X)$,
  - $E(aX + bY) = aE(X) + b\,E(Y)$,
  - $\mathrm{Var}(aX + bY) = a^2\mathrm{Var}(X) + b^2\mathrm{Var}(Y)$ for independent X and Y.

Before you start

You should know how to:

1. Calculate the mean and variance of a random variable

x	3	4	5
$P(X = x)$	0.1	0.6	0.3

   calculate $E(X)$ and $\mathrm{Var}(X)$.

   $E(X) = (3 \times 0.1) + (4 \times 0.6) + (5 \times 0.3) = 4.2$

   $E(X^2) = (9 \times 0.1) + (16 \times 0.6) + (25 \times 0.3)$
   $\qquad = 18$

   $\mathrm{Var}(X) = 18 - 4.2^2 = 0.36$

Skills check:

1. Calculate $E(X)$ and $\mathrm{Var}(X)$.

x	−2	1	5
$P(X = x)$	0.2	0.4	0.4

3.1 Expectation and variance of a linear function of a random variable

In S1, sections 5.3 and 5.4 you met the expectation and variance of a discrete random variable:

- The mean or expected value of a probability distribution is defined as
  $$\mu = E(X) = \sum px$$
- The variance of a probability distribution is defined as
  $Var(X) = E[\{X - E(X)\}^2]$. The alternative version (which is easier to use in practice) is $Var(X) = E(X^2) - \{E(X)\}^2$.

In S1, section 2.5 you saw:

If a set of data values X is related to a set of values Y so that $Y = aX + b$, then

- mean of $Y = a \times$ mean of $X + b$
- standard deviation of $Y = a \times$ standard deviation of X.
- variance of $Y = a^2 \times$ variance of X.

The same relationship applies if X and Y are random variables defined in the same way ($Y = aX + b$).

The proof of these results is easiest to do by considering the multiplication by a constant and adding a constant separately, and then the full result is obtained just by applying them one after the other. We will show it in full here for discrete random variables, but the same result holds for continuous random variables which you will meet in chapter 5 (where the summation is replaced by integration).

If $Y = aX$

$$\mu_X = \sum xp \Rightarrow \mu_Y = \sum yp = \sum (ax)\,p = a\sum xp = a\mu_X$$

$$E(X^2) = \sum x^2 p \Rightarrow E(Y^2) = \sum y^2 p = \sum (ax)^2\, p = a^2 \sum x^2 p = a^2 E(X^2)$$

$$Var(X) = E(X^2) - (\mu_X)^2;\ Var(Y) = E(Y^2) - (\mu_Y)^2 = a^2 E(X^2) - (a\mu_X)^2$$

$$= a^2\left\{E(X^2) - (\mu_X)^2\right\} = a^2\, Var(X)$$

If $Y = X + b$

$$\mu_X = \sum xp \Rightarrow \mu_Y = \sum yp = \sum (x + b)\, p = \sum xp + b\sum p = \mu_X + b \text{ (since } \sum p = 1)$$

It is now easier to use the other form for the variance i.e. $\text{Var}(X) = \text{E}(X - \mu_X)^2$ to derive the variance of Y:

$$\text{Var}(Y) = \text{E}(Y - \mu_Y)^2 = \text{E}((X + b) - (\mu_X + b))^2 = \text{E}(X - \mu_X)^2 = \text{Var}(X)$$

> Remember that variance is a measure of spread, so adding a constant to every piece of data, or to every possible outcome in a probability distribution simply shifts everything – it does not change how spread out it is.

For any random variable X:

$E(aX + b) = aE(X) + b$
$\text{Var}(aX + b) = a^2\text{Var}(X)$ where a and b are constants

---

Note that a and/or b can be negative and the algebra follows the same way – the square in the expression for variance ensures that the variance will always remain positive (if it is not zero).

Example 1

Given that $E(X) = 7$; $\text{Var}(X) = 3$ find:

a) i) $E(2X + 5)$ ii) $\text{Var}(2X + 5)$

b) i) $E(5 - 2X)$ ii) $\text{Var}(5 - 2X)$

· ·

a) i) $E(2X + 5) = 2E(X) + 5 = 14 + 5 = 19$

     ii) $\text{Var}(2X + 5) = 2^2\,\text{Var}(X) = 4 \times 3 = 12$

b) i) $E(5 - 2X) = 5 - 2E(X) = 5 - 14 = -9$

     ii) $\text{Var}(5 - 2X) = (-2)^2\,\text{Var}(X) = 4 \times 3 = 12$

Example 2

It costs $35 to hire a car for the day, and there is a mileage charge of 20 cents per mile. The distance travelled in a day has expectation 150 miles and standard deviation of 18 miles. Find the expectation and standard deviation of the cost per day.

· ·

Let X = number of miles travelled, and
Y = cost of hire in $ then $Y = 0.2X + 35$ ←— expressing the cost as a linear function of the mileage

$E(X) = 150$; $\text{Var}(X) = 18^2$

$E(Y) = 0.2 \times 150 + 35 = 65$; $\text{Var}(Y) = 0.2^2 \times 18^2 = (3.6)^2$

So the expectation of the cost per day is $65, and the standard deviation is $3.60.

Exercise 3.1

1. Given that $E(X) = 5.7$, $Var(X) = 1.9$ for each of the following functions, write down the mean and variance

    a) $2X + 7$ **b)** $4 - 3X$ **c)** $X + 3$ **d)** $7X$

2. **a)**

x	1	2	3	4	5
$P(X = x)$	0.1	0.2	0.3	0.3	0.1

    i) calculate $E(X)$ and $Var(X)$.

    ii) calculate the mean and variance of $5X - 2$.

    b)

y	-2	-1	0	1	2
$P(Y = y)$	0.1	0.2	0.3	0.3	0.1

    i) calculate $E(Y)$ and $Var(Y)$.

    ii) calculate the mean and variance of $5Y - 2$

    iii) $Y = X - 3$: show that $E(Y) = E(X) - 3$ and $Var(Y) = Var(X)$.

    c)

v	7	8	9	10	16
$P(V = v)$	$\frac{1}{2}$	$\frac{1}{4}$	$\frac{1}{8}$	$\frac{1}{16}$	$\frac{1}{16}$

    i) calculate $E(V)$ and $Var(V)$.

    ii) calculate the mean and variance of $4 + 3V$.

    d)

w	-2	-1	0	1	7
$P(W = w)$	$\frac{1}{2}$	$\frac{1}{4}$	$\frac{1}{8}$	$\frac{1}{16}$	$\frac{1}{16}$

    i) calculate $E(W)$ and $Var(W)$.

    ii) calculate the mean and variance of $4 + 3W$

    iii) $W = V - 9$: show that $E(W) = E(V) - 9$ and $Var(W) = Var(V)$.

3.

x	-2	-1	0	1	2
$P(X = x)$	0.1	0.4	0.2	0.2	0.1

   i) calculate $E(X)$ and $Var(X)$.

   ii) calculate the mean and variance of $7 - 2X$.

4. A supermarket sells top up vouchers valued at $5, $10, $15 or $20. The value, in dollars, of a top up voucher sold may be regarded as a random variable X with the following probability distribution:

x	5	10	15	20
$Pr(X = x)$	0.20	0.40	0.15	0.25

a) Find the mean and variance of X.

b) What is the probability that the next top up voucher sold is less than $15?

c) As a promotion the marketing department decides the supermarket will offer 10% off the cost of each top up voucher. Write down the mean and variance of the value of vouchers sold in the promotion, assuming customers continue to use the same pattern.

The marketing department had originally planned to offer $1 off each voucher.

d) i) What would be have been the mean and variance of the value of vouchers sold, assuming customers continued to use the same pattern.

ii) An experienced manager had told the supermarket that they would only sell $5 top up vouchers if they used this strategy. Explain why he thought this would happen.

5. A discrete random variable X has the probability function shown in the table below.

x	6	7	8	9
$P(X = x)$	0.1	0.2	0.3	0.4

Find

a) $P(X = 8)$,　　**b)** $E(X)$,　　**c)** $Var(X)$

d) $E(5 - 2X)$,　　**e)** $Var(5 - 2X)$

6. The random variable X has probability distribution

x	7	8	9	10	11
$P(X = x)$	0.2	a	0.3	0.1	b

a) Given that $E(X) = 9.05$, write down two equations involving a and b.

Find:

b) the value of a and the value of b,

c) $Var(X)$,

d) $Var(5 - 3X)$.

7. The random variable X has probability function

$$P(X = x) = \frac{(11 - 2x)}{25} \qquad x = 1, 2, 3, 4, 5.$$

a) Construct a table giving the probability distribution of X.

Find

b) $P(2 < X < 5)$　　**c)** $E(X)$

d) $Var(X)$　　**e)** $Var(2 - 3X)$.

8. The random variable X has the probability distribution:

x	10	12	15	16	18	20	24	25
$P(X \leq x)$	0.1	0.2	0.05	0.05	0.2	0.1	0.2	0.1

a) $E(X)$

b) $Var(X)$

c) $Var(20 - 3X)$.

9. An independent financial adviser recommends investments in 7 traded options. The number X from which the client makes a profit can be modelled by the discrete random variable with probability function.

$P(X = x) = kx$ $x = 0, 1, 2, 3, 4, 5, 6, 7$ where k is a constant.

a) Find the value of k.

b) Find $E(X)$ and $Var(X)$

The total cost of the investment is \$4000 and the return on each successful option is \$1500.

c) Find the probability that the client makes a loss overall.

d) Find the mean and variance of the profit the client makes.

10. A test is taken by a large number of members of the public. There are 5 questions, and the probability distribution of X, the number of correct answers given, is:

x	0	1	2	3	4	5
$P(X = x)$	0.05	0.10	0.25	0.30	0.20	0.10

a) Find the mean and standard deviation of X.

A mark, Y, is given where $Y = 10X + 5$.

b) Calculate the mean and standard deviation of Y.

3.2 Linear combination of two (or more) independent random variables

If X, Y are independent random variables then

$$E(aX \pm bY) = aE(X) \pm bE(Y)$$
$$\text{Var}(aX \pm bY) = a^2\,\text{Var}(X) + b^2\,\text{Var}(Y)$$

A formal proof of this is beyond the scope of this course (you would need to learn the theory of joint probability distributions to do it), but we can see examples of it in particular cases.

In fact, if we can show

$$E(X \pm Y) = E(X) \pm E(Y)$$
$$\text{Var}(X \pm Y) = \text{Var}(X) + \text{Var}(Y)$$

Then the general result will follow immediately using the results in the last section.

Consider the score on a fair die – it has probability distribution given by

x	1	2	3	4	5	6
p	$\frac{1}{6}$	$\frac{1}{6}$	$\frac{1}{6}$	$\frac{1}{6}$	$\frac{1}{6}$	$\frac{1}{6}$

Then $E(X) = 3.5$ and $\text{Var}(X) = \frac{35}{12}$. In S1, chapter 5 you met the distribution of the sum of the scores on two fair dice.

If S = sum of the scores on the two dice, the sample space for S will be:

S	1	2	3	4	5	6
1	2	3	4	5	6	7
2	3	4	5	6	7	8
3	4	5	6	7	8	9
4	5	6	7	8	9	10
5	6	7	8	9	10	11
6	7	8	9	10	11	12

In fact the result for the expectation holds even if X and Y are not independent, but at the moment we are only concerned with situations in which they are independent.

Then the probability distribution for S is:

s	2	3	4	5	6	7	8	9	10	11	12
$P(S = s)$	$\frac{1}{36}$	$\frac{2}{36}$	$\frac{3}{36}$	$\frac{4}{36}$	$\frac{5}{36}$	$\frac{6}{36}$	$\frac{5}{36}$	$\frac{4}{36}$	$\frac{3}{36}$	$\frac{2}{36}$	$\frac{1}{36}$

and the way the sample space has been coloured shows the pattern leading to the symmetrical triangular distribution.

Then $E(S) = 7$; $Var(S) = \dfrac{35}{6}$

So $E(S) = E(X) + E(Y)$; $Var(S) = Var(X) + Var(Y)$

If you now let D = difference between the two dice i.e. $D = X - Y$ where X and Y are the scores on two fair dice, you can see what happens

D	1	2	3	4	5	6
1	0	−1	−2	−3	−4	−5
2	1	0	−1	−2	−3	−4
3	2	1	0	−1	−2	−3
4	3	2	1	0	−1	−2
5	4	3	2	1	0	−1
6	5	4	3	2	1	0

There is a very similar pattern to the outcome space, but now the cells with the same value of D run along the other diagonal to those for the sum, and the mean and variance of D can be written down, looking at the distribution compared with that of S.

The probability distribution for D is:

d	−5	−4	−3	−2	−1	0	1	2	3	4	5
$P(D = d)$	$\frac{1}{36}$	$\frac{2}{36}$	$\frac{3}{36}$	$\frac{4}{36}$	$\frac{5}{36}$	$\frac{6}{36}$	$\frac{5}{36}$	$\frac{4}{36}$	$\frac{3}{36}$	$\frac{2}{36}$	$\frac{1}{36}$

$E(D) = 0 = E(X) - E(Y)$; $Var(D) = Var(S) = Var(X) + Var(Y)$

If we draw a graph showing the probability distributions you can see why the variance of the sum = the variance of the difference.

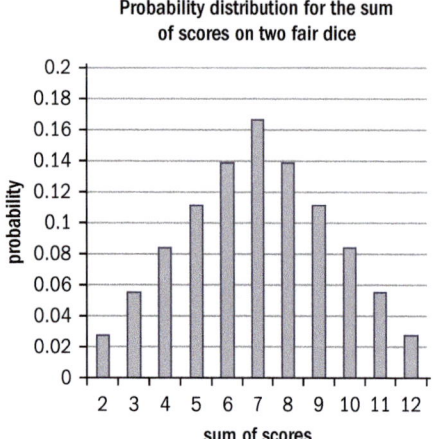

Probability distribution for the sum of scores on two fair dice

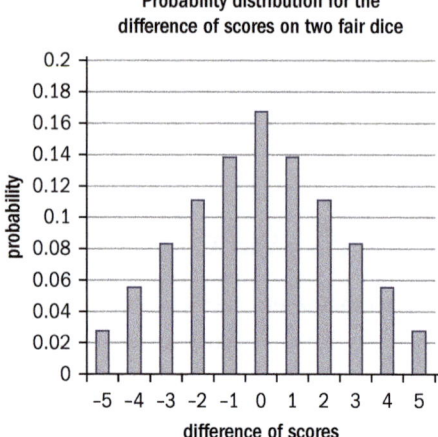

Probability distribution for the difference of scores on two fair dice

---

Example 3

If X is a random variable with $E(X) = 25$, $Var(X) = 8$, and Y is a random variable with $E(Y) = 15$, $Var(Y) = 6$, and X and Y are independent find:

i) $E(X + Y)$ and $Var(X + Y)$

ii) $E(X - Y)$ and $Var(X - Y)$

i) $E(X + Y) = E(X) + E(Y) = 25 + 15 = 40$,

    $Var(X + Y) = Var(X) + Var(Y) = 8 + 6 = 14$

ii) $E(X - Y) = E(X) - E(Y) = 25 - 15 = 10$

    $Var(X - Y) = Var(X) + Var(Y) = 8 + 6 = 14$

Example 4

A fair coin is tossed and a fair dice is thrown. Y is the score showing on the die, and $X = 0$ if the coin lands Heads and $X = 1$ is the coin lands Tails.

i)   Find the expectation and variance of each of X and Y,

ii) Find the probability distribution of $X + Y$.

iii) Use the probability distribution in ii) to find the expectation and variance of $X + Y$

iv) Show that $E(X + Y) = E(X) + E(Y)$ and $Var(X + Y) = Var(X) + Var(Y)$

i) X and Y are both equally likely distributions, given explicitly below:

x	0	1
$P(X = x)$	$\frac{1}{2}$	$\frac{1}{2}$

y	1	2	3	4	5	6
$P(Y = y)$	$\frac{1}{6}$	$\frac{1}{6}$	$\frac{1}{6}$	$\frac{1}{6}$	$\frac{1}{6}$	$\frac{1}{6}$

So $E(X) = \frac{1}{2}(0 + 1) = \frac{1}{2}$; $E(X^2) = \frac{1}{2}(0 + 1) = \frac{1}{2} \Rightarrow \text{Var}(X) = \frac{1}{2} - \left(\frac{1}{2}\right)^2 = \frac{1}{4}$

and $E(Y) = \frac{1}{6}(1 + 2 + 3 + 4 + 5 + 6) = \frac{7}{2}$; $E(Y^2) = \frac{1}{6}(1 + 4 + 9 + 16 + 25 + 36) = \frac{91}{6}$

$\Rightarrow \text{Var}(Y) = \frac{91}{6} - \left(\frac{7}{2}\right)^2 = \frac{35}{12}$

ii) Let $Z = X + Y$:

z	1	2	3	4	5	6	7
$P(Z = z)$	$\frac{1}{12}$	$\frac{1}{6}$	$\frac{1}{6}$	$\frac{1}{6}$	$\frac{1}{6}$	$\frac{1}{6}$	$\frac{1}{12}$

iii) $E(Z) = \frac{1}{6}(2 + 3 + 4 + 5 + 6) + \frac{1}{12}(1 + 7) = 4$;

$E(Z^2) = \frac{1}{6}(4 + 9 + 16 + 25 + 36) + \frac{1}{12}(1 + 49) = \frac{115}{6}$

$\Rightarrow \text{Var}(Z) = \frac{115}{6} - 4^2 = \frac{19}{6}$

iv) $E(X) + E(Y) = \frac{1}{2} + \frac{7}{2} = 4 = E(X + Y)$;

$\text{Var}(X) + \text{Var}(Y) = \frac{1}{4} + \frac{35}{12} = \frac{19}{6} = \text{Var}(X + Y)$

Exercise 3.2

1. If X is a random variable with $E(X) = 6$, $\text{Var}(X) = 5$, and Y is a random variable with $E(Y) = 10$, $\text{Var}(Y) = 2$, and X and Y are independent find:

   i) $E(X + Y)$ and $\text{Var}(X + Y)$

   ii) $E(X - Y)$ and $\text{Var}(X - Y)$

2. If X is a random variable with $E(X) = 15$, $\text{Var}(X) = 4$, and Y is a random variable with $E(Y) = 15$, $\text{Var}(Y) = 2$, and X and Y are independent find:

   i) $E(X + Y)$ and $\text{Var}(X + Y)$

   ii) $E(X - Y)$ and $\text{Var}(X - Y)$

3. If X is a random variable with $E(X) = 7$, $\text{Var}(X) = 1.5$, and Y is a random variable with $E(Y) = -4$, $\text{Var}(Y) = 2$, and X and Y are independent find:

   i) $E(X + Y)$ and $\text{Var}(X + Y)$

   ii) $E(X - Y)$ and $\text{Var}(X - Y)$

4. An examination consists of a written paper and an oral test. The written paper marks (A) have mean 68.6 and standard deviation 14.7. The oral test marks (B) are independent of the written paper marks and have mean 75.3 and standard deviation 6.2. The overall mark for the examination is found by adding 75% of A to 25% of B. Find the mean and standard deviation of the overall mark for the examination.

5. Camford has two branches of a bank. The number of customers on a Tuesday in the Middle Street branch has mean 45 and standard deviation 9. The number of customers on a Tuesday in the King's Street branch has mean 60 and standard deviation 12.

   i) Assuming the number of customers in the two branches are independent of one another, find the mean and standard deviation of the total number of customers using the two branches on a Tuesday.

   ii) Comment on whether the assumption made in part i) is reasonable.

6. A fair coin is tossed and C = number of heads seen. A fair dice is thrown and D = score seen on the dice. The random variable X is defined as $X = D - 2C$.

   i) Find the mean and variance of C.

   ii) Given that $E(D) = 3.5$ and $Var(D) = \dfrac{35}{12}$, find the mean and variance of X.

3.3 Expectation and variance of a sum of repeated independent observations of a random variable, and the mean of those observations

Applying the results in the last two sections gives some results that are really important to the ideas of estimation and hypothesis testing you will meet in later chapters.

If $X_1, X_2, X_3, \ldots\ldots X_n$ are independent observations of a random variable X with $E(X) = \mu$, $Var(X) = \sigma^2$ then:

$$E\left(\sum_{i=1}^{n} X_i\right) = n\mu, \quad Var\left(\sum_{i=1}^{n} X_i\right) = n\sigma^2 \text{ and}$$

$$E(\overline{X}) = E\left(\frac{1}{n}\sum_{i=1}^{n} X_i\right) = \frac{n\mu}{n} = \mu, \quad Var(\overline{X}) = Var\left(\frac{1}{n}\sum_{i=1}^{n} X_i\right) = \frac{n\sigma^2}{n^2} = \frac{\sigma^2}{n}$$

The last of these four results is very important in statistics because it tells us that the sample mean is less variable than the individual observations, and that the variance decreases as the sample size increases. The sample mean becomes increasingly likely to be close to the population mean as the sample size increases.

Example 5

X is a random variable with $E(X) = 24$, $Var(X) = 8$. Find the expectation and variance of

i) $\displaystyle\sum_{i=1}^{5} X_i$

ii) $\displaystyle \bar{X}_{12} = \frac{1}{12}\sum_{i=1}^{12} X_i$

· ·

i) $Y = \displaystyle\sum_{i=1}^{5} X_i \Rightarrow E(Y) = 5 \times E(X) = 120; Var(Y) = 5 \times Var(X) = 40$

ii) $E(\bar{X}_{12}) = E(X) = 24; \quad Var(\bar{X}_{12}) = \dfrac{Var(X)}{12} = \dfrac{8}{12} = \dfrac{2}{3}$

Example 6

X is a random variable with $E(X) = 17.2$, $Var(X) = 45$.

Find the smallest sample size for which the standard deviation of the sample mean will be not greater than 2.

· ·

$Var(\bar{X}_n) = \dfrac{45}{n} \le 2^2 \Rightarrow 4n \ge 45 \Rightarrow n \ge 11.25$ so the smallest sample size to

meet this criterion is 12.

Exercise 3.3

1. X is a random variable with $E(X) = 15$, $Var(X) = 4$. Find the expectation and variance of

   i) $\displaystyle\sum_{i=1}^{6} X_i$

   ii) $\displaystyle \bar{X}_6 = \frac{1}{6}\sum_{i=1}^{6} X_i$

2. X is a random variable with $E(X) = 0$, $Var(X) = 20$. Find the expectation and variance of

   i) $\displaystyle\sum_{i=1}^{20} X_i$

   ii) $\displaystyle \bar{X}_{20} = \frac{1}{20}\sum_{i=1}^{20} X_i$

3. X is a random variable with $E(X) = -5$, $Var(X) = 25$. Find the expectation and variance of

   i) $\displaystyle\sum_{i=1}^{50} X_i$

   ii) $\displaystyle \bar{X}_{50} = \frac{1}{50}\sum_{i=1}^{50} X_i$

4. X is a random variable with $E(X) = 35.2$, $Var(X) = 26.1$. Find the expectation and variance of

   i) $\displaystyle\sum_{i=1}^{10} X_i$

   ii) $\displaystyle \bar{X}_{10} = \frac{1}{10}\sum_{i=1}^{10} X_i$

5. X is a random variable with $E(X) = -5.2$, $Var(X) = 21$.

   Find the smallest sample size for which the standard deviation of the sample mean will be not greater than 2.

6. X is a random variable with $E(X) = 522.1$, $Var(X) = 39.2$. Find the smallest sample size for which the standard deviation of the sample mean will be not greater than 1.

3.4 Comparing the sum of repeated independent observations with the multiple of a single observation*

There is no easy way to derive the probability distribution of the sum of a number of dice thrown together, but it is relatively easy to simulate the situation and compare what you see when a large number of repetitions are taken. The graphs below show 500 sums of 10 dice, and 500 single throws multiplied by 10:

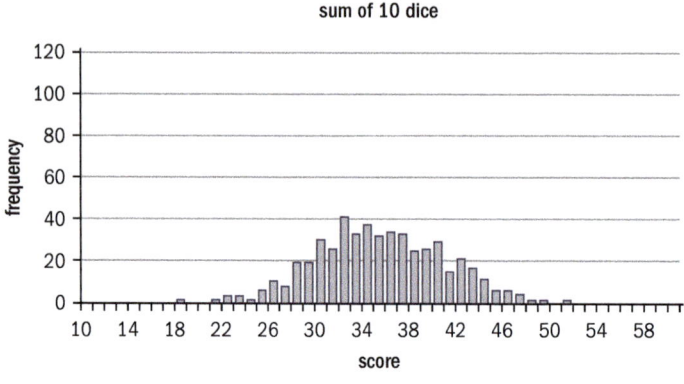

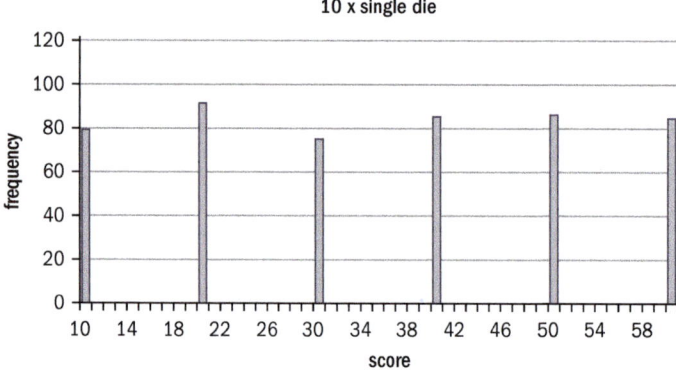

The graphs have the axes kept to identical scales to enable the comparison to be made.

Only six outcomes are possible in the second case – where a single observation has been multiplied by 10 – and those six outcomes are equally likely, although the 500 observations are not exactly evenly spread. When the sum of ten independent throws of a dice are taken, any (integer) score from 10 to 60 could be observed, but to get a total score of 10 you would need to see a 1 on each of the ten dice, which would be extremely unusual – not impossible, but occurring less than one in sixty million times that ten dice are thrown. In contrast, a single one occurs on average once in six throws, and in the experimental results shown above there were 80 occasions in 500 trials for which a score of 10 was obtained.

The outcomes for the sum have a lot more possible values, but those near the middle of the possible range have a very high proportion of all the observations recorded (474 of the 500 observations were between 25 and 45 in this case), and the distribution is visibly much less spread out than the lower graph showing the single throw multiplied by 10.

Note: There is no exercise for this section as it is not within the syllabus.

Summary exercise 3

1. Given that $E(X) = 12.1$, $Var(X) = 2.4$ for each of the following functions, write down the mean and variance
   a) $2X + 3$
   b) $5 - 3X$
   c) $X - 3$
   d) $9X$

2.

x	12	13	14	15	16
$P(X = x)$	0.1	0.2	0.4	0.2	0.1

   i) calculate $E(X)$ and $Var(X)$.
   ii) calculate the mean and variance of $4X - 3$

3. The random variable X has probability distribution

x	7	8	9	10	11
$P(X = x)$	0.2	a	0.3	0.1	b

   a) Given that $E(X) = 9.2$, write down two equations involving a and b.

   Find:
   b) the value of a and the value of b,
   c) $Var(X)$,
   d) $Var(3 - 2X)$.

4. The random variable X has probability function
$$P(X = x) = \frac{(9 - 2x)}{16} \qquad x = 1, 2, 3, 4.$$
   a) Construct a table giving the probability distribution of X.

   Find
   b) $P(2 \leq X \leq 4)$
   c) $E(X)$
   d) $Var(X)$
   e) $Var(2 + 3X)$.

5. If X is a random variable with $E(X) = 10$, $Var(X) = 3$, and Y is a random variable with $E(Y) = 7$, $Var(Y) = 3$, and X and Y are independent find:
   i) $E(X + Y)$ and $Var(X + Y)$
   ii) $E(X - Y)$ and $Var(X - Y)$

6. An examination consists of a written paper and an oral test. The written paper marks (A) have mean 53.9 and standard deviation 8.9. The oral test marks (B) are independent of the written paper marks and have mean 64.5 and standard deviation 6.8. The overall mark for the examination is found by adding 50% of A to 50% of B. Find the mean and standard deviation of the overall mark for the examination.

7. The discrete random variable X has probability function $P(X = x) = kx^2$ for $x = 1, 2, 3, 4$ where k is a positive constant.
   a) Show that $k = \dfrac{1}{30}$. [2]
   b) Find $E(X)$ and show that $E(X^2) = 11.8$. [4]
   c) Find $Var(X)$. [1]
   d) Find the mean and variance of $Y = 17 - 4X$.

8. A discrete random variable X has a probability function as shown in the table below, where a and b are constants.

x	1	2	4	8
$P(X = x)$	0.2	0.3	a	b

Given that $E(X) = 4.4$,

a) find the value of a and the value of b. [5]

Find

b) $P(1 < X < 6)$, [1]

c) $E(3X - 4)$. [2]

d) Show that $Var(X) = 9.24$. [3]

e) Evaluate $Var(3X - 4)$.

EXAM-STYLE QUESTION

9. Exam marks, X, have mean 64.5 and standard deviation 12.5. The marks need to be scaled using the formula $Y = aX + b$ so that the scaled marks, Y, have mean 70 and standard deviation 10. Find the values of a and b.

10. A fair coin is tossed 5 times and the number of heads is recorded.

i) The random variable X is the number of heads. State the mean and variance of X. [2]

ii) The number of heads is doubled and denoted by the random variable Y. State the mean and variance of Y. [2]

Cambridge International A Level and AS Mathematics 9709 Paper 3 Q1 May/June 2003

11. The independent random variables X and Y are such that X has mean 8 and variance 4.8 and Y has a Poission distribution with mean 6. Find

i) $E(2X - 3Y)$. [2]

ii) $Var(2X - 3Y)$. [4]

Cambridge International A Level and AS Mathematics 9709 Paper 7 Q3 May/June 2004

Chapter summary

- For any random variable X and constants a, b:

    $\mathrm{E}(aX + b) = a\mathrm{E}(X) + b$
    $\mathrm{Var}\,(aX + b) = a^2\mathrm{Var}(X)$

- If X, Y are independent random variables then

    $\mathrm{E}(aX + bY) = a\mathrm{E}(X) + b\,\mathrm{E}(Y)$
    $\mathrm{Var}(aX + bY) = a^2\mathrm{Var}(X) + b^2\mathrm{Var}(Y)$

- If $X_1, X_2, X_3, \ldots\ldots X_n$ are independent observations of a random variable X with $\mathrm{E}(X) = \mu, \quad \mathrm{Var}(X) = \sigma^2$ then:

    $$\mathrm{E}\left(\sum_{i=1}^{n} X_i\right) = n\mu, \quad \mathrm{Var}\left(\sum_{i=1}^{n} X_i\right) = n\sigma^2 \text{ and}$$

    $$\mathrm{E}(\bar{X}) = \mathrm{E}\left(\frac{1}{n}\sum_{i=1}^{n} X_i\right) = \frac{n\mu}{n} = \mu, \; \mathrm{Var}(\bar{X}) = \mathrm{Var}\left(\frac{1}{n}\sum_{i=1}^{n} X_i\right) = \frac{n\sigma^2}{n^2} = \frac{\sigma^2}{n}$$

- Remember two special cases:
    - If X, Y are independent random variables then

        $\mathrm{E}(X - Y) = \mathrm{E}(X) - \mathrm{E}(Y)$
        $\mathrm{Var}\,(X - Y) = \mathrm{Var}\,(X) + \mathrm{Var}\,(Y)$

    - $X_1, X_2, X_3, \ldots\ldots X_n$ are independent observations of a random variable X then:

        $$\mathrm{E}\left(\sum_{i=1}^{n} X_i\right) = n\mu = n\mathrm{E}(X); \; \mathrm{Var}\left(\sum_{i=1}^{n} X_i\right) = n\sigma^2 \neq \mathrm{Var}\,(nX) = n^2\sigma^2$$

Review exercise A

1. Exam marks, X, have mean 70 and standard deviation 8.7. The marks need to be scaled using the formula $Y = aX + b$ so that the scaled marks, Y, have mean 55 and standard deviation 6.96. Find the values of a and b. **[4]**

Cambridge International AS and A Level Mathematics 9709 Paper 7 Q1 June 2005

2. A shopkeeper sells electric fans. The demand for fans follows a Poisson distribution with mean 3.2 per week.
   i) Find the probability that the demand is exactly 2 fans in any one week. **[2]**
   ii) The shopkeeper has 4 fans in his shop at the beginning of a week. Find the probability that this will not be enough to satisfy the demand for fans in that week. **[4]**
   iii) Given instead that he has n fans in his shop at the beginning of a week, find, by trial and error, the least value of n for which the probability of his not being able to satisfy the demand for fans in that week is less than 0.05. **[4]**

Cambridge International AS and A Level Mathematics 9709 Paper 7 Q6 November 2005

3. The number of radioactive particles emitted per second by a certain metal is random and has mean 1.7. The radioactive metal is placed next to an object which independently emits particles at random such that the mean number of particles emitted per second is 0.6. Find the probability that the total number of particles emitted in the next 3 seconds is 6, 7 or 8. **[4]**

Cambridge International AS and A Level Mathematics 9709 Paper 7 Q1 November 2004

4. People arrive randomly and independently at the elevator in a block of flats at an average rate of 4 people every 5 minutes.

   i) Find the probability that exactly two people arrive in a 1-minute period. **[2]**
   ii) Find the probability that nobody arrives in a 15-second period. **[2]**
   iii) The probability that at least one person arrives in the next t minutes is 0.9. Find the value of t. **[4]**

Cambridge International AS and A Level Mathematics 9709 Paper 7 Q6 June 2008

5. Of people who wear contact lenses, 1 in 1500 on average have laser treatment for short sight.
   i) Use a suitable approximation to find the probability that, of a random sample of 2700 contact lens wearers, more than 2 people have laser treatment. **[4]**
   ii) In a random sample of n contact lens wearers the probability that no one has laser treatment is less than 0.01. Find the least possible value of n. **[3]**

Cambridge International AS and A Level Mathematics 9709 Paper 7 Q5 November 2004

6. The random variable X denotes the number of worms on a one metre length of a country path after heavy rain. It is given that X has a Poisson distribution.
   i) For one particular path, the probability that $X = 2$ is three times the probability that $X = 4$. Find the probability that there are more than 3 worms on a 3.5 metre length of this path. **[5]**
   ii) For another path the mean of X is 1.3.
      a) On this path the probability that there is at least 1 worm on a length of k metres is 0.96. Find k. **[4]**
      b) Find the probability that there are more than 1250 worms on a one kilometre length of this path. **[3]**

Cambridge International AS and A Level Mathematics 9709 Paper 7 Q6 November 2007

7. The random variable X has the distribution Po(1.3). The random variable Y is defined by $Y = 2X$.

   i) Find the mean and variance of Y. [3]

   ii) Give a reason why the variable Y does not have a Poisson distribution. [1]

   Cambridge International AS and A Level Mathematics 9709 Paper 72 Q1 November 2011

8. The cost of hiring a bicycle consists of a fixed charge of 500 cents together with a charge of 3 cents per minute. The number of minutes for which people hire a bicycle has mean 142 and standard deviation 35.

   i) Find the mean and standard deviation of the amount people pay when hiring a bicycle. [3]

   ii) 6 people hire bicycles independently. Find the mean and standard deviation of the total amount paid by all 6 people. [3]

   Cambridge International AS and A Level Mathematics 9709 Paper 71 Q3 November 2012

9. A dressmaker makes dresses for Easifit Fashions. Each dress requires $2.5\,\text{m}^2$ of material. Faults occur randomly in the material at an average rate of 4.8 per $20\,\text{m}^2$.

   i) Find the probability that a randomly chosen dress contains at least 2 faults. [3]

   Each dress has a belt attached to it to make an outfit. Independently of faults in the material, the probability that a belt is faulty is 0.03. Find the probability that, in an outfit,

   ii) neither the dress nor its belt is faulty, [2]

   iii) the dress has at least one fault and its belt is faulty. [2]

   The dressmaker attaches 300 randomly chosen belts to 300 randomly chosen dresses. An outfit in which the dress has at least one fault and its belt is faulty is rejected.

   iv) Use a suitable approximation to find the probability that fewer than 3 outfits are rejected. [3]

   Cambridge International AS and A Level Mathematics 9709 Paper 7 Q6 June 2006

10. Major avalanches can be regarded as randomly occurring events. They occur at a uniform average rate of 8 per year.

    i) Find the probability that more than 3 major avalanches occur in a 3-month period. [3]

    ii) Find the probability that any two separate 4-month periods have a total of 7 major avalanches. [3]

    iii) Find the probability that a total of fewer than 137 major avalanches occur in a 20-year period. [4]

    Cambridge International AS and A Level Mathematics 9709 Paper 71 Q3 June 2009

11. In restaurant A an average of 2.2% of tablecloths are stained and, independently, in restaurant B an average of 5.8% of tablecloths are stained.

    i) Random samples of 55 tablecloths are taken from each restaurant. Use a suitable Poisson approximation to find the probability that a total of more than 2 tablecloths are stained. [4]

    ii) Random samples of n tablecloths are taken from each restaurant. The probability that at least one tablecloth is stained is greater than 0.99. Find the least possible value of n. [4]

    Cambridge International AS and A Level Mathematics 9709 Paper 71 Q6 June 2010

12. An examination consists of a written paper ana a practical test. The written paper marks (M) have mean 54.8 and standard deviation 16.0. The practical test marks (P) are independent of the written paper marks and have mean 82.4 and standard deviation 4.8. The final mark is found by adding 75% of M to 25% of P. find the mean and standard deviation of the final marks for the examination. [3]

    Cambridge International AS and A Level Mathematics 9709 Paper 71 Q2 June 2012

Maths in real-life

The mathematics of the past

In Pure 2/3 chapter 2 you met the phenomenon of radioactive decay and the behavior that half of the remaining radioactive atoms in any substance will decay in a period known as the half-life of the substance. The rate at which atoms decay follows (roughly) a Poisson process with the average rate in a fixed period reducing over time. Later in this book you will meet the Central Limit Theorem which is one of a number of instances you will meet in statistics where the long term behavior of random events is very stable (often known as *The Laws of Large Numbers*) – and the predictability of the half-life characteristic is an example of a practical situation in which observed reality matches the theoretical predictions.

In the middle of the last century, Willard Libby developed a method of dating organic materials discovered in archeological excavations. The method relies on the naturally occurring carbon isotope ^{14}C (Carbon 14) which is created by the action of cosmic ray neutrons on Nitrogen 14 atoms in the upper atmosphere. The ^{14}C atoms quickly oxidise to form carbon dioxide in the air. All organic materials (plants, animals, humans) assimilate ^{14}C from the atmosphere throughout their lifetimes. However, when they die, they stop exchanging carbon with the atmosphere and it then starts to decay at the known rate.

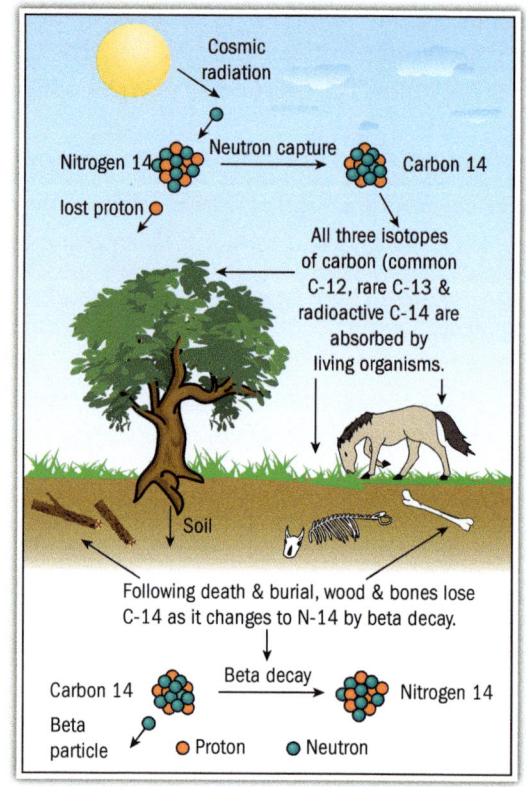

The fact that the proportion of ^{14}C amongst all carbon elements in the atmosphere is known means that measuring the proportion now allows a dating process. Libby's work is generally regarded as having one of the widest impacts of any scientific discovery in the 1900s and he was awarded the Nobel Prize for Chemistry in 1960, but his work revolutionised archaeology and has applications in other human and physical sciences.

When the technique was introduced the calculations were based on a Geiger counter detecting the number of decaying particles, but other more accurate methods were soon developed and now the technique is commonly based on using *Accelerometer Mass Spectrometry* which determines the proportions of ^{14}C atoms and the

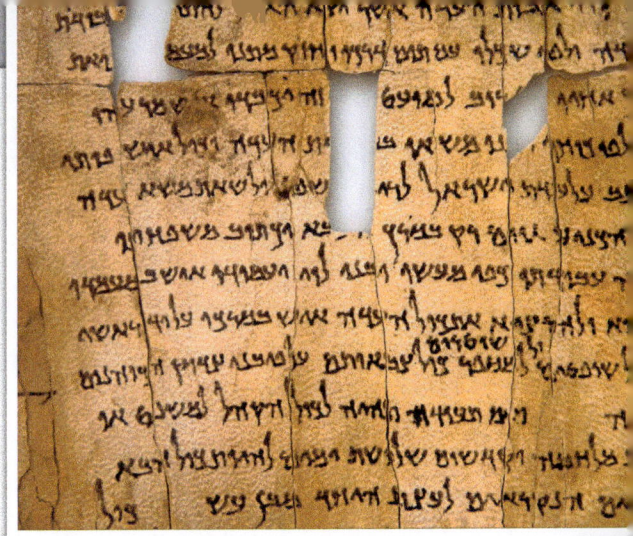

other carbon isotopes. Calculations for dates using different detection methods for ^{14}C atoms use different mathematics, but all require the use of a base line of what the ^{14}C presence has been historically. Initially this was thought to be constant but work analysing tree ring data series has given more accurate calibration data extending back some 14,000 years and several other variations have since been detected so different calibration scales are used in northern and southern hemispheres, and for marine organisms.

The Dead Sea Scrolls are a collection of texts which were discovered in caves about 2 kilometres from the northwest shore of the Dead Sea. The texts are of enormous historical and religious significance and carbon dating suggests that the majority of the scrolls belong to the last two centuries BC and the first century AD.

Newspapers will tend to sensationalise research findings and attribute precision to them which the scientists do not claim – headlines might use language like 'Carbon dating pinpoints Mayan calendar' when reporting the analysis of an ancient beam from a Guatemalan temple which the scientists concluded came from a tree which was cut down and carved around AD 658 – 696.

The process of carbon dating is heavily statistical – dates are reported using interval estimates giving an idea of both when the artifact was created and some idea of the precision with which the date is known (you will meet interval estimates formally in chapter 7 of this book). Current developments on the technique of carbon dating focus on improving the calibration scales used for identifying the best estimate for the date (the centre of the range reported) and on how to improve the reliability of the estimates – which would allow narrower ranges to be reported.

The Step Pyramid of Djoser in Saqqara was shown to be the oldest stone pyramid in Egypt by carbon dating techniques.

Manufacturers need to closely control production levels – it there is less chocolate in a bar or box than the label says they can be prosecuted, but if there is consistently more than the label says it costs the company money. The bars shown are made from a single mould, so there is one distribution to control, but the situation in a box of chocolates is that it depends on the sum of a number of distributions, so they need to be able to control what happens to the sum by knowing how it depends on the individual distributions.

Objectives

After studying this chapter you should be able to use, in the course of solving problems, the results that:

- if X has a normal distribution then so does $aX + b$,
- if X and Y have independent normal distributions then $aX + bY$ has a normal distribution,
- if X and Y have independent Poisson distributions then $X + Y$ has a Poisson distribution.

Before you start

You should know how to:

1. Calculate probabilities using the Poisson distribution. X: Po(6) Find $P(X = 2)$

$$P(X = 2) = \frac{e^{-6} 6^2}{2!} = 0.0446$$

2. Calculate probabilities using the Normal distribution

$X \sim N(10, 25)$ Find $P(X < 0)$

$$P(X < 0) = P\left(Z < \frac{0 - 10}{\sqrt{25}} = -2 \right) = 0.0228$$

Skills check:

1. $X \sim Po(2.1)$ Find $P(X = 1)$

2. $X \sim N(20, 3.4)$ Find $P(X > 21.1)$

4.1 The distribution of the sum of two independent Poisson random variables

In chapter 3 you saw that the expectation and variance of the sum of **any** two independent random variables are given by the sum of the expectations and the sum of the variances of the two random variables, but this tells you nothing about the probability distribution of the sum. You also saw the detail of the distribution of the sum of two uniform distributions (the scores on two dice) which gave a triangular distribution, so you know that it is not generally the case that the distribution of the sum will be in a similar form to the original.

However, there are two special cases where the sum does have the same form. The first of these is the Poisson distribution:

> The sum of two independent Poisson random variables is still a Poisson:
> i.e. $P(\lambda) + P(\mu) \sim P(\lambda + \mu)$

The full proof of this requires some sophisticated algebraic manipulation, in particular using the binomial expansion, and it is noted at the end of this section.

There are a couple of things you can consider which may help with why this distribution might have the special regenerative property.

- Remember that for a Poisson (λ) random variable the mean and variance are the same, $E(X) = \lambda$, $\text{Var}(X) = \sigma^2 = \lambda$, and so the general results of the expectation and variance of the sum will mean that, whatever the detailed distribution is, it will have the property that the mean and variance will be equal.

- Remember that changing the length of interval being considered for a Poisson distribution simply resulted in a proportional change to the parameter to be used – so if the number of telephone calls arriving at a switchboard in a ten minute period follows a Poisson distribution and, on average, there are 7 calls in that time, then the number of calls in a 5 minute period would follow a P(3.5) distribution, and the number of calls in a 20 minute period would follow a P(14) distribution. You could think of the twenty minute period as the total in two separate 10 minute intervals.

The key steps in the full proof are much easier to see in a concrete example. Consider the probability that $X + Y$ takes the value 3 when $X \sim P(2.3)$ and $Y \sim P(3.8)$ and X, Y are independent. For the sum to be 3, the possible combinations for the values of X and Y are 0,3; 1,2; 2,1 and 3,0 and you can calculate that probability:

$$P(X + Y = 3) = \left(e^{-2.3} \times \frac{3.8^3 e^{-3.8}}{3!}\right) + \left(\frac{2.3^1 e^{-2.3}}{1!} \times \frac{3.8^2 e^{-3.8}}{2!}\right) + \left(\frac{2.3^2 e^{-2.3}}{2!} \times \frac{3.8^1 e^{-3.8}}{1!}\right) + \left(\frac{2.3^3 e^{-2.3}}{3!} \times e^{-3.8}\right)$$

$$= e^{-2.3} \times e^{-3.8}\left(\frac{3.8^3}{3!} + \frac{3.8^2 2.3^1}{2!1!} + \frac{3.8^1 2.3^2}{1!2!} + \frac{2.3^3}{3!}\right)$$

$$= e^{-2.3} \times e^{-3.8}\left(\frac{(3.8+2.3)^3}{3!}\right) = \frac{6.1^3 e^{-6.1}}{3!}$$

which is the probability of getting a 3 for a P(6.1) distribution.

Proof that if $X \sim P(\lambda)$ and $Y \sim P(\mu)$ and X, Y are independent, then $X + Y \sim P(\lambda + \mu)$:

$$\Pr\{X+Y=k\} = \sum_{i=0}^{k}\Pr\{X=i \text{ and } Y = k-i\} = \sum_{i=0}^{k}\frac{e^{-\lambda}\lambda^i}{i!} \times \frac{e^{-\mu}\mu^{k-i}}{(k-i)!}$$

$$= \frac{e^{-\lambda} \times e^{-\mu}}{k!}\sum_{i=0}^{k}\frac{k!}{i!(k-i)!} \times \lambda^i \mu^{k-i} = \frac{e^{-\lambda} \times e^{-\mu}}{k!}(\lambda + \mu)^k$$

Example 1

Two radioactive substances emit, on average, 3 electrons per second and 2 electrons per second respectively.

Calculate the probability that a total of exactly 4 electrons are emitted in a given second.

· ·

If X is the total number of electrons emitted in a second, then $X \sim P(5)$.

And $\Pr(X = 4) = \dfrac{e^{-5}5^4}{4!} = 0.175$ (3 s.f.)

Example 2

The number of goals a team scores in a league match may be modelled by a Poisson distribution with mean 2.2. The number of goals the same team concedes in a league match may be modelled by a Poisson distribution with mean 1.5.

a) Assuming these are independent of one another, find the probability that:
   i) the match finishes as a 1–1 draw.
   ii) there are two goals in the match
b) Comment on the assumption that the number of goals scored and conceded by a team are independent of one another.

· ·

a) i) Need both random variables to take the value 1, so
$$\Pr(\text{score is } 1\text{–}1) = \frac{e^{-2.2}2.2^1}{1!} \times \frac{e^{-1.5}1.5^1}{1!} = 0.24377 \times 0.334695 = 0.0816 \text{ (3 s.f.)}$$
   ii) You could work out the other two possible score (2–0 and 0–2) and add them to the answer in part i), or use the distribution for the total goals is P(3.7).
$$\Pr(\text{two goals}) = \frac{e^{-3.7}3.7^2}{2!} = 0.169 \text{ (3 s.f.)}$$
b) This is unlikely to be true because it is a competitive situation where the difference between 3–1 and 2–1 is not the same as the score going to 2–2 (a draw) from 2–1.

Exercise 4.1

1. The demand for two magazines in a newsagents form (independent) Poisson distributions. On average the monthly demand for Internet Investigator is 4.3, and for the Web Wanderer it is 2.9. Calculate the probability that:

   a) the newsagent is asked for 3 Internet Investigators and 3 Web Wanderers in one month;

   b) he is asked for 5 magazines altogether in one month;

   c) he is asked for 5 magazines in two months of one year.

2. **a)** Serious accidents in a certain type of manufacturing industry can be adequately modelled by the Poisson distribution with a mean rate of 1.6 per week.

   i) What is the probability that there are no serious accidents in a particular week?

   ii) What is the probability that there are at least three serious accidents in a three week period?

   iii) What is the probability that in a four week period there is exactly one week in which there are serious accidents?

   b) Minor accidents in this manufacturing industry can also be adequately modelled by the Poisson distribution with a mean rate of 5.4 per week, and these can be assumed to occur independently of the serious accidents.

   i) State the distribution for the total number of accidents in the industry.

   ii) What is the probability that there are at least five accidents in a particular week?

3. Two radioactive substances emit, on average, 3.2 electrons per second and 1.3 electrons per second respectively.

   a) Calculate the probability that a total of exactly 3 electrons are emitted in a given second.

   b) Calculate the probability that a total of at least 7 electrons are emitted in a two second period.

4. You are given that $X \sim \text{Po}(5)$ and $Y \sim \text{Po}(3)$ are independent random variables. Which of the following have a Poisson distribution? State the parameter value for any which are Poisson variables.

   i) $3X + 2$ **ii)** $2X + 3Y$ **iii)** $X + Y$ **iv)** $X - Y$

5. The number of vehicles passing a point on a motorway heading east has a Poisson distribution with mean 24 per minute. The number of vehicles passing the same point on the motorway heading west has a Poisson distribution with mean 21 per minute. Find the probability that the total number of vehicles passing the point in a 10-second period is less than 5.

6. The number of customers entering a jewellery shop in the first hour that it opens can be modelled by a Poisson distribution with mean 2.4, and the number of customers entering for the rest of the day can be modelled by a Poisson with mean 3.1 per hour. The shop opens at 0900. Find the probability that fewer than 4 customers enter the shop between 0930 and 1200.

4.2 Linear functions and combinations of normal random variables

In chapter 3 you saw:

- the relationships between the expectation and variance of a linear function of X and the expectation and variance of X;
- the expectation and variance of the sum of **any** two independent random variables are given by the sum of the expectations and the sum of the variances of the two random variables;
- the expectation and variance of the difference of **any** two independent random variables are given by the difference of the expectations and the sum of the variances of the two random variables.

The normal distribution is the second instance of where the distribution of the sum is in the same form as the original distributions, but it is also true in this case for the difference (which it is not the case for the Poisson), and for a linear function of a single normal random variable (which again is not the case for the Poisson). Using the results from chapter 3 we can deduce what the parameters will be:

If $X \sim N(\mu_X, \sigma_X^2)$ and $Y \sim N(\mu_Y, \sigma_Y^2)$ are independent then:

- $aX + b \sim N(a\mu_X + b, a^2\sigma_X^2)$

- $aX + bY \sim N(a\mu_X + b\mu_Y, a^2\sigma_X^2 + b^2\sigma_Y^2)$

The full proof of this requires the use of much more sophisticated statistical methods (moment generating functions), and unlike the Poisson there are not any simple examples you can do to satisfy yourself that it works in particular cases.

Example 3

A container is known to have mass 40 grams. The amount of liquid a machine dispenses into the container follows a normal distribution with mean 200 ml and standard deviation 10 ml.
The liquid has density 0.85 kg per litre.
Find the probability that the filled container weighs less than 200 grams.

· ·

If X = volume of liquid in ml, and Y = mass of filled container in grams then $Y = 40 + 0.85X$.

An alternative method would be to work out the volume of liquid which would give the total mass of 200 grams ($= \dfrac{160}{0.85} = 188.2$ ml) and use the distribution of the volume of liquid dispensed.

Given that $X \sim N(200, 10^2)$ this means

$$Y \sim N(40 + 0.85 \times 200, 0.85^2 \times 10^2) \Rightarrow Y \sim N(210, 8.5^2)$$

$$\Pr\{Y < 200\} = \Pr\left\{Z < \frac{200-210}{8.5} = -1.176\right\} = \Phi(-1.176)$$

$$= 1 - \Phi(1.176) = 1 - 0.8802 = 0.120$$

So there is a 12% chance that the filled container weighs less than 200 grams.

In example 3 the container was known to have a mass of 40 grams. More realistically on a production line like this, the containers would also have a distribution which can be modelled by a normal distribution. The next example asks the same question as in example 3, but using a distribution for the mass of the container.

> ## Example 4
>
> A container has mass which is normally distributed with mean 40 grams, and standard deviation 3 grams. The amount of liquid a machine dispenses into the container follows a normal distribution with mean 200 ml and standard deviation 10 ml.
> The liquid has density 0.85 kg per litre. Find the probability that the filled container weighs less than 200 grams.
>
> ···
>
> If X = volume of liquid in ml, Y = mass of empty container in grams and W = mass of the filled container in grams then $W = 0.85X + Y$.
>
> Given that $X \sim N(200, 10^2)$ and $Y \sim N(40, 3^2)$ this means
>
> $$W \sim N(0.85 \times 200 + 40, 0.85^2 \times 10^2 + 3^2) \Rightarrow Y = N(210, 81.25)$$
>
> $$Pr = \{W < 200\} = Pr \left\{ Z < \frac{200 - 210}{\sqrt{81.25}} = -1.109 \right\} = \Phi(-1.109)$$
>
> $$= 1 - \Phi(1.109) = 1 - 0.866 = 0.134$$
>
> So there is now a 13.4% chance that the filled container weighs less than 200 grams.

Note that in this case the alternative method noted for example 3 is not possible because you don't know the mass of either the container or the liquid.

There are two important special cases to take particular note of:

- The mean of n independent normal random variables is still a normal with $\bar{X}_n \sim N\left(\mu_X, \dfrac{\sigma_X^2}{n}\right)$

- The difference of two independent normal random variables is still a normal with $N(\mu_X, \sigma_X^2) - N(\mu_Y, \sigma_Y^2) \sim N(\mu_X - \mu_Y, \sigma_X^2 + \sigma_Y^2)$

The first of these results will be considered in much more depth later – in chapter 6 on sampling and chapter 7 on estimation, and it is the foundation for testing the mean of a normal distribution which you will meet in chapter 9.

The second result is important when you consider situations in which the relative size of two measurements is more important that their actual sizes:

- in an engine, a piston moves up and down inside a cylinder – obviously the piston diameter has to be smaller than the cylinder for it to fit, but it is also important that it is not too much smaller: if it is then the piston will wobble around and cause wear on the sides of the cylinder, and also not be as efficient as some of the energy can escape past the sides of the piston rather than driving the engine.

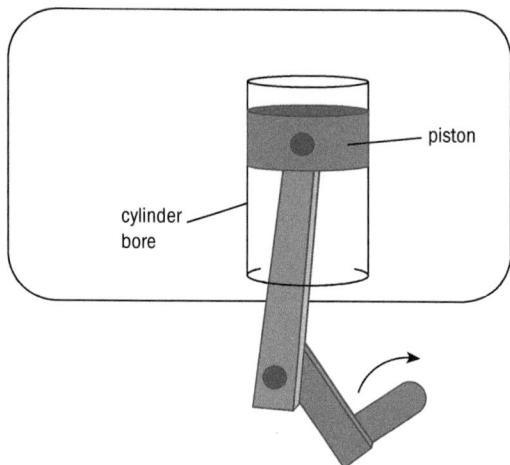

- A similar situation applies to a nut and bolt – and again there will be a tolerance for how much smaller the bolt diameter can be than the nut diameter – otherwise the threads will not catch and the bolt can slip through the nut.

- Which of two athletes wins in a competition? The times or distances recorded by an athlete in an event can often be (approximately) modelled by a normal distribution. The sign of the difference in their times or distances determines who wins. However, the result depends on the independence of the two distributions, which is unlikely to be exactly true in a head to head competition if the athletes can see where they are relative to the other athlete during the competition. It may still be a useful first approximation – in statistics you are often faced with situations in which the only thing you can do is something you know is not a perfect description of the situation – as long as you treat the result with caution, knowing its imperfections, it is likely to be better than having no information to work with.

Example 5

A batch of bolts has diameters which are normally distributed with mean 17.5 mm and standard deviation 0.4 mm. The diameters of holes of the batch of nuts delivered with the bolts are normally distributed with mean 18 mm and standard deviation 0.3 mm,

i) Find the probability that a randomly chosen bolt will fit inside a randomly chosen nut.

If the diameter of the bolt is more than 1.1 mm smaller than the diameter of the hole then the bolt is not securely held.

ii) Find the probability that a randomly chosen bolt which fits inside a randomly chosen nut will be held securely.

· ·

If X = diameter of the bolt, and Y = diameter of the nut then $X \sim N(17.5, 0.4^2)$ and $Y \sim N(18, 0.3^2)$.

i) $Y - X \sim N(0.5, 0.3^2 + 0.4^2) = N(0.5, 0.5^2)$

For the bolt to fit inside the nut, you need $Y - X > 0$

$$\Pr\{Y - X > 0\} = \Pr\left\{Z > \frac{0 - 0.5}{0.5} = -1\right\} = 1 - \Phi(-1) = \Phi(1) = 0.8413$$

ii) This is a conditional probability – that it holds securely **given that** it fitted inside. If it fits inside and the bolt is held securely then $0 < Y - X < 1.1$

$$\Pr\{0 < Y - X < 1.1\} = \Pr\left\{-1 < Z < \frac{1.1 - 0.5}{0.5} = 1.2\right\}$$
$$= \Phi(1.2) - \Phi(-1) = \Phi(1.2) - (1 - \Phi(1))$$
$$= 0.8849 - (1 - 0.8413) = 0.7262$$

$$\Pr\{0 < Y - X < 1.1 | Y - X > 0\} = \frac{0.7272}{0.8413} = 0.863$$

Exercise 4.2

1. You are given that $X \sim N(5, 7)$ and $Y \sim N(3, 10)$ are independent random variables. Which of the following have a Normal distribution? State the parameters for any which are Normally distributed

    i) $3X + 2$ **ii)** $2X + 3Y$ **iii)** $X + Y$ **iv)** $X - Y$

2. $X \sim N(320, 20)$ and $Y \sim N(110, 30)$ are independent random variables.

    i) State the distribution of $2X + 3Y$

    ii) Find the probability that $2X + 3Y$ is more than 1000.

3. $X \sim N(40, 4.5)$ and $Y \sim N(62.5, 12.1)$ are independent random variables.

    i) State the distribution of $5X + 4Y$

    ii) Find the probability that $5X + 4Y$ is more than 440.

4. $A \sim N(12, 7)$ and $B \sim N(-3, 2)$ are independent random variables.

    i) State the distribution of $4A - 3B$

    ii) Find the probability that $4A - 3B$ is more than 55.

5. The cost of Jarinda's electricity for a month is a fixed charge of 800 cents together with a charge of 5.8 cents per unit of electricity used. The number of units she uses in a month is normally distributed with mean 456 units and standard deviation 18.2 units.

    i) Find the mean and standard deviation of the total cost of Jarinda's electricity in a randomly chosen month.

    ii) Find the probability that her bill in a randomly chosen month is more than 35 dollars.

6. W is the mass, in grams, of water in a bottle. B is the mass, in grams, of the bottle. $W \sim N(752, 8)$ and $B \sim N(30, 1.2)$ are independent.

    i) Find the probability that a filled bottle has a total mass of less than 775 grams.

    ii) A litre of water weighs 1 kilogram. Find the probability that 4 randomly chosen bottles contain less than 3 litres of water.

7. The bottles of water described in question 6 are packed in cardboard boxes which contain 24 bottles. The mass C, in grams, of the cardboard box is Normally distributed with mean 750 grams and standard deviation 6.2 grams.

    a) i) Find the distribution of the mass of a box with 24 filled bottles.

        ii) Find the probability that the full box has a mass of more than 19.5 kg.

    b) A shop takes delivery of 20 full boxes. Find the probability that the average mass of the boxes is more than 19.5 kg.

Summary exercise 4

1. **a)** Minor accidents in a particular factory can be adequately modelled by the Poisson distribution with a mean rate of 3.6 per week.

    i) What is the probability that there are no minor accidents in a particular week?

    ii) What is the probability that there are at least five serious accidents in a three week period?

    ii) What is the probability that in a four week period there is exactly three weeks in which there are minor accidents?

    b) Serious accidents in this factory can also be adequately modelled by the Poisson distribution with a mean rate of 0.3 per week, and these can be assumed to occur independently of the minor accidents.

    i) State the distribution for the total number of accidents in the industry.

    ii) What is the probability that there are at least five accidents in a particular week?

2. The number of vehicles passing a point on a rural road heading north has a Poisson distribution with mean 3.5 per hour. The number of vehicles passing the same point on the road heading south has a Poisson distribution with mean 3.2 per hour.

    i) Find the probability that no vehicles pass the point in a randomly chosen 15 minute period.

    ii) Find the probability that the total number of vehicles passing the point in a two-hour period is less than 6.

3. The cost of Grigor's gas for a month is a fixed charge of 500 cents together with a charge of 6.4 cents per unit of gas used. The number of units he uses in a month is normally distributed with mean 965 units and standard deviation 83.2 units.

    i) Find the mean and standard deviation of the total cost of Grigor's gas in a randomly chosen month.

    ii) Find the probability that his bill in a randomly chosen month is more than 65 dollars.

4. W is the mass, in grams, of water in a bottle. B is the mass, in grams, of the bottle. $W \sim N(501, 5.2)$ and $B \sim N(25, 1.5)$ are independent.

    i) Find the probability that a filled bottle has a total mass of less than 525 grams.

    ii) A litre of water weighs 1 kilogram. Find the probability that 4 randomly chosen bottles contain more than 2 litres of water.

5. The volume of water in bottles made by Aguafresh is normally distributed with mean 751 millilitres and standard deviation 9.2 millilitres. The volume of water in bottles made by Lifespring is normally distributed with mean 1002 millilitres and standard deviation 12.8 millilitres. Find the probability that 4 randomly chosen bottles of Aguafresh contain less liquid than 3 randomly chosen bottles of Lifespring.

6. In their hockey matches, the Steelers score goals independently and at random times with an average of 2.8 goals per match.

i) State the expected number of goals that the Steelers will score in the first half of a match.

ii) Find the probability that the Steelers will score twice in the first half of a match and not in the second half.

iii) Given that the Steelers score two goals in a match, find the conditional probability that both goals were scored in the first half.

iv) Hockey matches last for 70 minutes. In a particular match the Steelers score one goal in the first ten minutes. Find the probability that they will score at least two goals in the match.

Independently of the number of goals scored by the Steelers, the number of goals scored per hockey match by the Giants has a Poisson distribution with mean 1.7.

v) Find the probability that more than 3 goals will be scored in a particular match when the Steelers play the Giants.

7. The number of characters on a page of a book can be modelled by a normal distribution with mean 1983 and standard deviation 44.7. Find the probability that the total number of characters in a random sample of 8 pages is more than 16 000.

8. The weights of men follow a normal distribution with mean 74 kg and standard deviation 7.5 kg. The weights of women follow a normal distribution with mean 60.5 kg and standard deviation 5.4 kg.

Four randomly selected men sit on one side of a large see-saw and five randomly selected women sit on the other side. What is the probability that the men's side is heavier than the women's side?

9. The weights of pebbles on a beach are normally distributed with mean 48.5 grams and standard deviation 12.4 grams.

i) Find the probability that the mean weight of a random sample of 5 pebbles is greater than 51 grams. [3]

ii) The probability that the mean weight of a random sample of n pebbles is less than 51.6 grams is 0.9332. Find the value of n. [4]

Cambridge International A Level and AS Mathematics 9709 Paper 71 Q3 November 2009

10. Weights of cups have a normal distribution with mean 91 g and standard deviation 3.2 g. Weights of saucers have an independent normal distribution with mean 72 g and standard deviation 2.6 g. Cups and saucers are chosen at random to be packed in boxes, with 6 cups and 6 saucers in each box. Given that each empty box weights 550 g, find the probability that the total weight of a box containing 6 cups and 6 saucers exceeds 1550 g. [5]

Cambridge International A Level and AS Mathematics 9709 Paper 71 Q3 May/June 2003

11. Kieran and Andreas are long-jumpers. They model the lengths, in metres, that they jump by the independent random variables $K \sim N(5.64, 0.0576)$ and $A \sim N(4.97, 0.0441)$ respectively. They each make a jump and measure the length. Find the probability that

i) the sum of the lengths of their jumps is less than 11 m, [4]

ii) Kieran jumps more than 1.2 times as far as Andreas. [6]

Cambridge International A Level and AS Mathematics 9709 Paper 71 Q7 November 2013

12. A journey in a certain car consists of two stages with a stop for filling up with fuel after the first stage. The length of time, T minutes, taken for each stage has a normal distribution with mean 74 and standard

deviation 7.3. The length of time, F minutes, it takes to fill up with fuel has a normal distribution with mean 5 and standard deviation 1.7. The length of time it takes to pay for the fuel is exactly 4 minutes. The variables T and F are independent and the times for the two stages are independent of each other.

i) Find the probability that the total time for the journey is less than 154 minutes. [5]

ii) A second car has a fuel tank with exactly twice the capacity of the first car. Find the mean and variance of this car's fuel fill-up time. [2]

iii) This second car's time for each stage of the journey follows a normal distribution with mean 69 minutes and standard deviation 5.2 minutes. The length of time it takes to pay for the fuel for this car is also exactly 4 minutes. Find the probability that the total time for the journey taken by the first car is more than the total time taken by the second car. [5]

Cambridge International A Level and AS Mathematics 9709 Paper 7 Q7 November 2005

13. At work Jerry receives emails randomly at a constant average rate of 15 emails per hour.

i) Find the probability that Jerry receives more than 2 emails during a 20-minute period at work. [3]

ii) Jerry's working day is 8 hours long. Find the probability that Jerry receives fewer than 110 emails per day on each of 2 working days. [4]

iii) At work Jerry also receives texts randomly and independently at a constant average rate of 1 text every 10 minutes. Find the probability that the total number of emails and texts that Jerry receives during a 5-minute period at work is more than 2 and less than 6. [4]

Cambridge International A Level and AS Mathematics 9709 Paper 71 Q7 May/June 2012

14. In athletics matches the triple jump event consists of a hop, followed by a step, followed by a jump. The lengths covered by Albert in each part are independent normal variables with means 3.5 m, 2.9 m, 3.1 m and standard deviations 0.3 m, 0.25 m, 0.35 m respectively. The length of the triple jump is the sum of the three parts.

i) Find the mean and standard deviation of the length of Albert's triple jumps. [3]

ii) Find the probability that the mean of Albert's next four triple jumps is greater than 9 m. [3]

Cambridge International A Level and AS Mathematics 9709 Paper 7 Q2 May/June 2004

Chapter summary

- The sum of two independent Poisson random variables is still a Poisson:
  If $X \sim \text{Po}(\lambda)$ and $Y \sim \text{Po}(\lambda)$ are independent then $X + Y \sim \text{Po}(\lambda + \mu)$

- The sum of two independent Normal random variables is still a Normal:
  If $X \sim N\left(\mu_X, \sigma_X^2\right)$ and $Y \sim N\left(\mu_Y, \sigma_Y^2\right)$ are independent then:

  $$aX + b \sim N\left(a\mu_X + b, a^2\sigma_X^2\right)$$

  $$aX + bY \sim N\left(a\mu_X + b\mu_Y, a^2\sigma_X^2 + b^2\sigma_Y^2\right)$$

5 Continuous random variables

Since any random variable which is a measurement of time, length etc. will be a continuous random variable, the material in this chapter is fundamental to the realistic treatment of much of the world around us. The management of risk is now very big business – modelling the behaviour of tides, earthquakes, floods etc. is important both for insurance companies and for governments in determining policy such as whether to allow residential developments in areas of potential flood risk.

Objectives

After studying this chapter you should be able to:

- Understand the concept of a continuous random variable, and recall and use properties of a probability density function (restricted to functions defined over a single interval);
- use a probability density function to solve problems involving probabilities, and to calculate the mean and variance of a distribution (explicit knowledge of the cumulative distribution function is not included, but location of the median, for example, in simple cases by direct consideration of an area may be required).

Before you start

You should know how to:

1. Integrate a variety of functions, and use the results. If $\int_0^{\frac{\pi}{4}} k \sin x \, dx = 1$ find the value of k.

$$\int_0^{\frac{\pi}{4}} k \sin x \, dx = [-k \cos x]_0^{\frac{\pi}{4}} = k\left(-\frac{1}{\sqrt{2}} - -1\right)$$

$$= k\left(\frac{\sqrt{2}-1}{\sqrt{2}}\right) = 1 \Rightarrow k = \frac{\sqrt{2}}{\sqrt{2}-1}$$

$$\int_0^{\frac{\pi}{4}} k \sin x \, dx = [-k \cos x]_0^{\frac{\pi}{4}} = k\left(-\frac{1}{\sqrt{2}} - -1\right)$$

$$= k\left(\frac{\sqrt{2}-1}{\sqrt{2}}\right) = 1 \Rightarrow k = \frac{\sqrt{2}}{\sqrt{2}-1}$$

Skills check:

1. If $\int_0^1 (2 - kx) \, dx = 1$ find the value of k.

2. Calculate $\int_{-1}^2 (x^2 - 2x + 5) \, dx$

5.1 Introduction to continuous random variables

The times taken by competitors in the swimming section of a triathlon are shown in the following histograms, with smaller and smaller intervals:

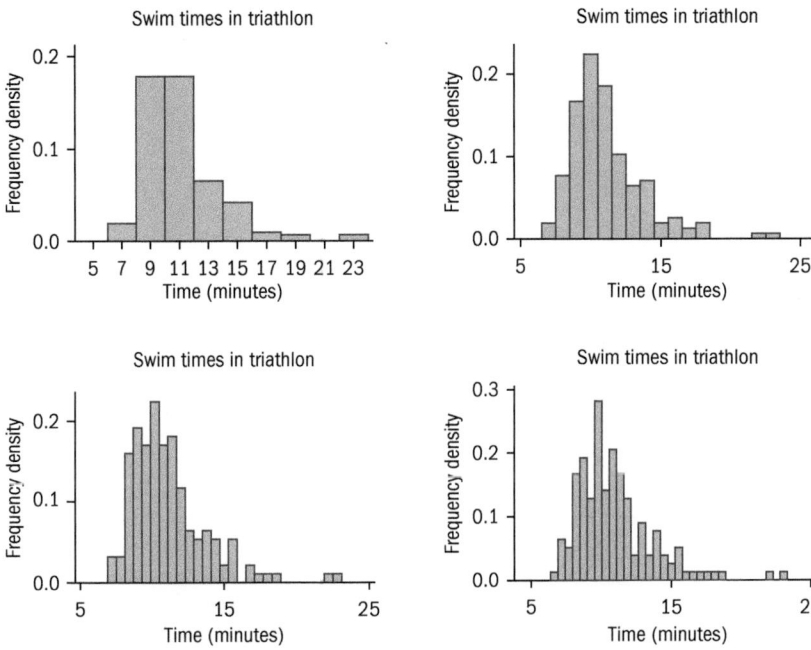

As the number of intervals increases, the picture of the data shows jagged edges. If this process was continued, taking smaller and smaller intervals, it looks like the following graph is a good description of the distribution of times for the swim section of the triathlon in that competition.

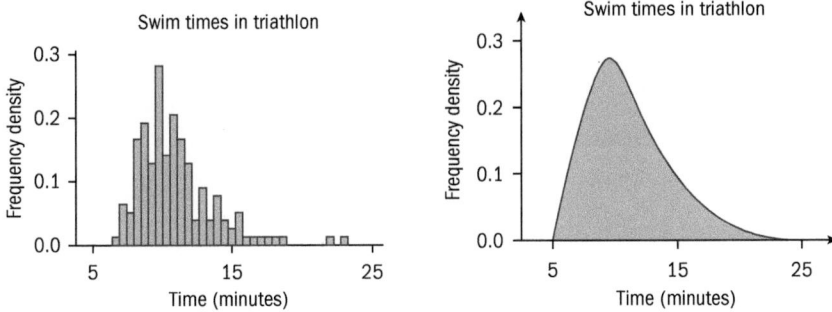

On a different day, when the weather is different, when the competitors are different, or on a different course, the distribution might be different.

These histograms are drawn as **frequency density** diagrams, so the total area of all the bars is always 1.

The limit of the cases where smaller and smaller bars are taken, but the total area stays at 1 is what is meant by a **continuous random variable**.

The probabilities are defined as the area under the curve between two values of x.

Measurements of length, time, mass, area and volume as well as compound measures such as speed and density are on a continuous scale. While they are often recorded by being grouped in intervals, the actual distribution is really continuous.

> The Normal distribution is one of the most commonly occurring continuous random variables. The times for the swimming in the triathlon do not have the symmetry of the Normal distribution.

Exercise 5.1

1. State which of the following random variables are continuous:

    a) The length of the winning throw in the men's javelin competition at the next Olympic Games.

    b) The number of blades of grass in a 1 square metre area of lawn.

    c) The volume of water in a bottle.

    d) The length of time it takes Habib to walk to school on a particular morning.

    e) The number of times Habib is late for school in a particular term.

2. The histogram shows the durations of the eruptions of the Old Faithful Geyser in Yellowstone Park over a period of time.

    Sketch what you think the graph of the continuous random variable looks like.

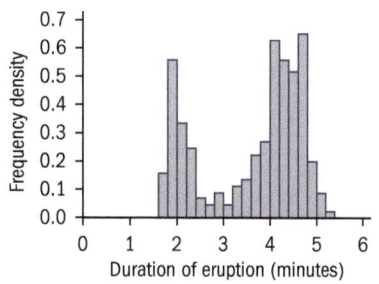

3. The histogram shows the time between eruptions of the Old Faithful Geyser.

    Sketch what you think the graph of the continuous random variable looks like.

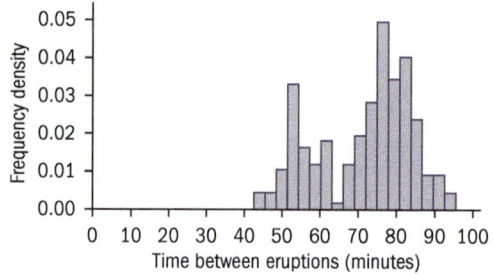

4. The histogram shows the weights in grams of a sample of African grey parrots.

Sketch what you think the graph of the continuous random variable looks like.

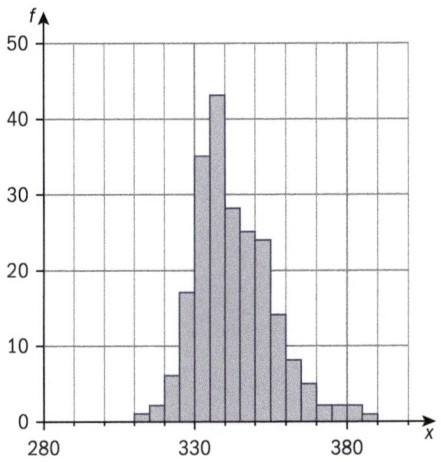

5.2 Probability density functions

To be a **probability density function (pdf)**, $f(x)$ must satisfy these basic properties:

- $f(x) \geq 0$ for all x so that all probabilities are not negative

- $\int_{-\infty}^{\infty} f(x) \cdot dx = 1$ – often $f(x)$ is only defined over a small range, in which case the integral over that range will be 1.

You can find the probability that a random variable lies between $x = a$ and $x = b$ from the area under the curve represented by $f(x)$ between those two points.

Example 1

Show that $f(x)$ is a probability density function where

$f(x) = \frac{1}{2}(x - 3)$ for $3 \leq x \leq 5$.

Find P($X < 4$).

For a continuous random variable, there is no difference between $<$ and $\leq$ (or between $>$ and $\geq$).

$f(x) \geq 0$ for all x it is defined for.

$$\int_{-\infty}^{\infty} f(x) \cdot dx = \int_{3}^{5} \left[\frac{1}{2}(x - 3)\right] \cdot dx = \left[\frac{1}{4}x^2 - \frac{3}{2}x\right]_{3}^{5}$$

$$= \left(\frac{25}{4} - \frac{15}{2}\right) - \left(\frac{9}{4} - \frac{9}{2}\right) = 1$$

so $f(x)$ is a probability density function.

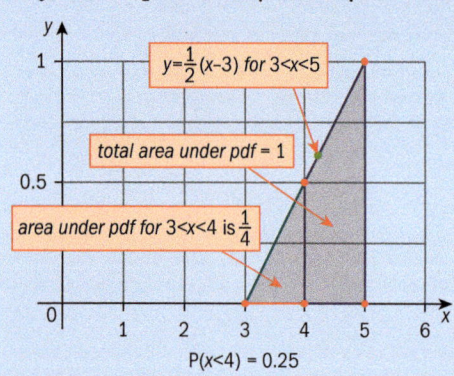

Example 2

For the following functions, state whether they could be used as a pdf, and if not explain why.

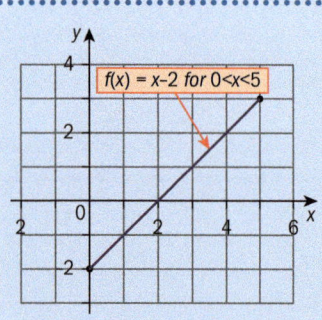

No – because there are negative values for $f(x)$

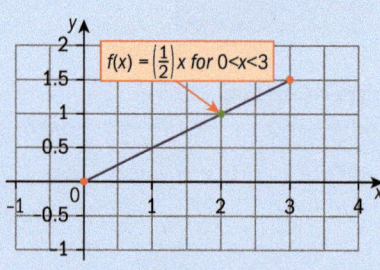

No because the area under the pdf is 2.25

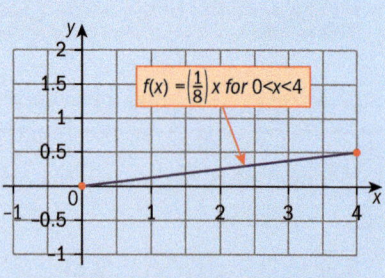

Yes – there are no negative values and the total area = 1

You can find the area under a curve $y = f(x)$ between $x = a$ and $x = b$ by calculating the integral $\int_{b}^{a} f(x) \cdot dx$.

See Pure 2 for revision.

If $f(x)$ is a **probability density function (pdf)**, then
$$P(a < X < b) = \int_a^b f(x) \cdot dx.$$

Example 3

$f(x) = k(9 - x^2)$ for $-3 \le x \le 3$

Find **a)** the value of k and calculate $P(-1 < X < 2)$

      b) $P(X = 2)$

a) $\int_{-3}^{3} k(9 - x^2)\,dx = k\left[9x - \dfrac{x^3}{3}\right]_{-3}^{3} = k((27 - 9) - (-27 - -9)) = 36k$

so $k = \dfrac{1}{36}$.

$P(-1 < X < 2) = \int_{-1}^{2} \dfrac{1}{36}(9 - x^2)\,dx = \dfrac{1}{36}\left[9x - \dfrac{x^3}{3}\right]_{-1}^{2}$

$= \dfrac{1}{36}\left(\left(18 - \dfrac{8}{3}\right) - \left(-9 - \dfrac{-1}{3}\right)\right) = \dfrac{24}{36} = \dfrac{2}{3}$

b) For a continuous distribution the probability of a single value is 0.

If you are given a sketch of a pdf, you might be required to write down the function form of the pdf in simple cases.

Example 4

Express the pdf shown in this sketch as a function.

Find the equation of the line through $A(2,0)$ and $B(3, 2)$:

The gradient is $m = \dfrac{2-0}{3-2} = 2$,

so the equation is $y = 2x + c$

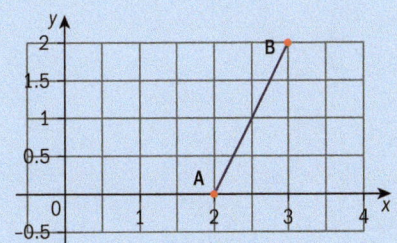

Substitute $(2, 0)$: $0 = 4 + c$, so $c = -4$.

Equation is $y = 2x - 4$.

The full form of the pdf is then
$f(x) = 2x - 4$ for $2 \le x \le 3$ (and $f(x) = 0$ elsewhere).

You know that the median of a data set is a value which has half the data less than or equal to it, and the same is true for probability distributions, so α is the median of a distribution if $P(X \le \alpha) = 0.5$.

In general this would be done by integration of the pdf using an upper limit of α and setting the value equal to 0.5 and solving the resulting equation for α, but in this course you are only expected to find the median in cases where it can be found by the direct consideration of an area.

Example 5

Find the median for the probability density function in example 1
$f(x) = \frac{1}{2}(x - 3)$ for $3 \leq x \leq 5$.

If α is the median then the sides of the triangle shown are $(\alpha - 3)$ and $\frac{1}{2}(\alpha - 3)$, so using the formula for the area of a triangle gives

$$\frac{1}{2}(\alpha - 3)\frac{1}{2}(\alpha - 3) = \frac{1}{2} \Rightarrow (\alpha - 3)^2 = 2 \Rightarrow \alpha = 3 + \sqrt{2}.$$

Alternative solution:

$$\int_3^\alpha \left\{\frac{1}{2}(x - 3)\right\} \cdot dx = \left[\frac{1}{4}x^2 - \frac{3}{2}x\right]_3^\alpha = 0.5$$

$$\Rightarrow \left(\frac{\alpha^2}{4} - \frac{3\alpha}{2}\right) - \left(\frac{9}{4} - \frac{9}{2}\right) = 0.5$$

$$\Rightarrow \alpha^2 - 6\alpha + 7 = 0$$

$$\Rightarrow \alpha = \frac{6 \pm \sqrt{36 - 28}}{2} = 3 \pm \sqrt{2}$$

$$\Rightarrow \alpha = 3 + \sqrt{2}$$

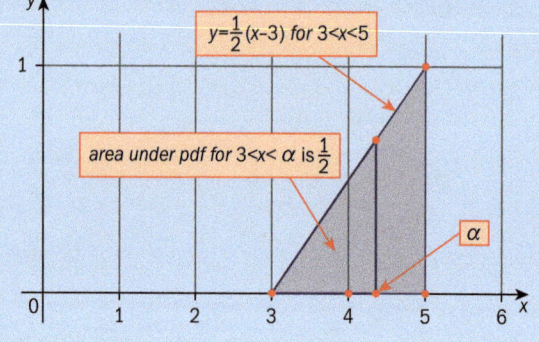

$y = \frac{1}{2}(x-3)$ for $3 < x < 5$

area under pdf for $3 < x < \alpha$ is $\frac{1}{2}$

Exercise 5.2

1. Which of the following functions could be probability density functions?

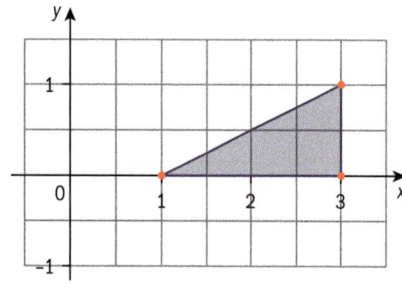

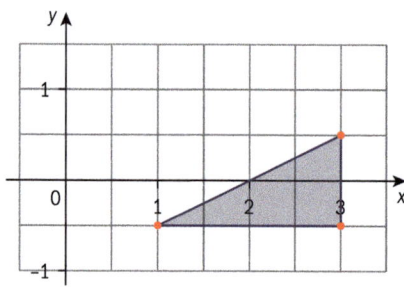

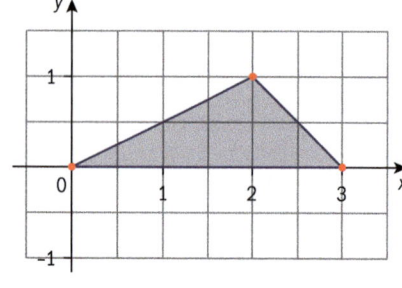

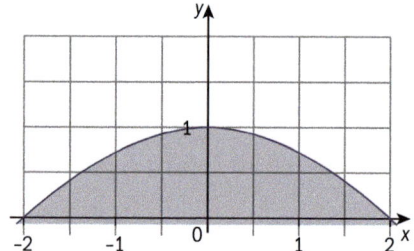

2. The graphs of two probability density functions are shown.

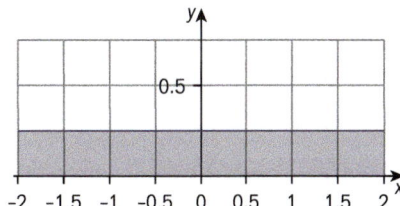 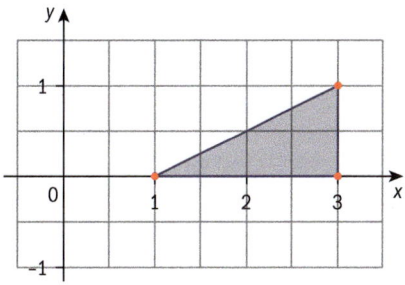

a) Find **i)** $P(X < 2)$ **ii)** $P(|X| < 1)$ **b)** Find **i)** $P(X < 2)$ **ii)** $P(X < 2.5)$

3. Find the value of k for which the following functions can represent a probability density function (in each case $f(x)$ is 0 outside the defined range).

a) $f(x) = kx^2$ $-1 \leq x \leq 1$ **b)** $f(x) = kx$ $0 \leq x \leq 3$

c) $f(x) = \frac{x}{6} + k$ $0 \leq x \leq 3$ **d)** $f(x) = kx^2$ $0 \leq x \leq 1$

e) $f(x) = \frac{x}{4}$ $1 \leq x \leq k$

4. The following functions represent a probability density function (in each case $f(x)$ is 0 outside the defined range).

a) $f(x) = \frac{x}{2}$ $0 \leq x \leq 2$ Find $P(X < 1)$

b) $f(x) = \frac{1}{8}x$ $0 \leq x \leq 4$ Find $P(1 < X < 2)$

c) $f(x) = \frac{x}{10} + k$ $0 \leq x \leq 2$ Find k and $P(X > 1)$

d) $f(x) = \frac{3}{8}x^2$ $0 \leq x \leq 2$

Find **i)** $P(X < 1)$ **ii)** $P(X = 1.5)$

e) $f(x) = kx^2$ $1 \leq x \leq 3$ Find k and $P(X < 2.5)$

5. For the following probability density functions, give the equation.
Make sure you define it for all values of x from $-\infty$ to ∞.

a) 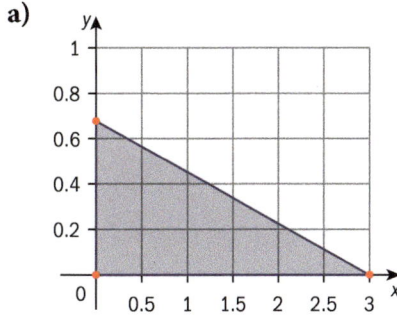 **b)**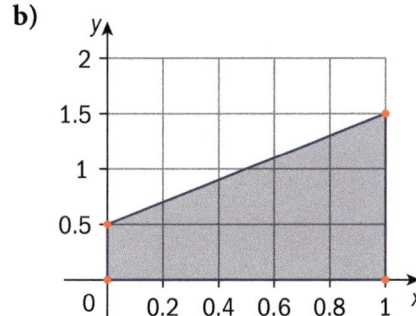

Note: In part **a)** the intercept is at $\left(0, \frac{2}{3}\right)$

6. For the probability density functions in the corresponding parts of question 5, find

   a) **i)** $P(X > 2)$ **ii)** $P(X < 0.5)$

   b) **i)** $P(X > 0.2)$ **ii)** $P(X < 1)$

5.3 Mean and variance of a continuous random variable

The **mean** or **expected value** of a discrete probability distribution is defined as

$$\mu = E(X) = \sum px$$

See Statistics 1, Chapter 5 for revision.

For a continuous random variable

$$\mu = E(X) = \int_{-\infty}^{\infty} xf(x)\cdot dx$$

The pdf has replaced p and the summation has become an integral.

where in practice the limits will be the limits of the interval over which $f(x)$ is defined.

Example 6

$f(x) = \frac{1}{2}(x - 3)$ for $3 \le x \le 5$.   Find $E(X)$.

Use the formula:

$$\mu = E(X) = \int_{-\infty}^{\infty} xf(x)\cdot dx$$

This is $f(x)$ from Example 1

$$= \int_{3}^{5} \left\{ x \cdot \frac{1}{2}(x - 3) \right\} \cdot dx$$

$$= \int_{3}^{5} \left\{ \frac{x^2}{2} - \frac{3x}{2} \right\} \cdot dx$$

$$= \left[\frac{1}{6}x^3 - \frac{3}{4}x^2 \right]_{3}^{5}$$

$$= \left(\frac{125}{6} - \frac{75}{4} \right) - \left(\frac{27}{6} - \frac{27}{4} \right) = \frac{13}{3}$$

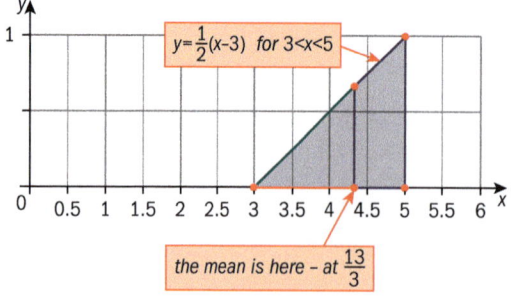

$y = \frac{1}{2}(x-3)$ for $3 < x < 5$

the mean is here – at $\frac{13}{3}$

If you balanced this triangle on a pivot at the mean $\left(4\frac{1}{3}\right)$ it would stay in equilibrium – there is only $\frac{4}{9}$ of the area to the left, but it extends further left than the larger area extends to the right of the mean.

Recall the definition of variance:
$$\text{Var}(X) = \sigma^2 = \text{E}[\{X - \text{E}(X)\}^2]$$

You almost always use the computational form
$\sigma^2 = \text{E}(X^2) - \mu^2$ where $\mu = \text{E}(X)$.

For a continuous random variable $\text{E}(X^2) = \displaystyle\int_{-\infty}^{\infty} x^2\, f(x) \cdot dx$

Example 7

Find the standard deviation for the pdf in Example 1.

$$\text{E}(X^2) = \int_{-\infty}^{\infty} x^2 f(x) \cdot dx$$

$$= \int_{3}^{5} \left\{ x^2 \cdot \frac{1}{2}(x - 3) \right\} \cdot dx$$

$$= \int_{3}^{5} \left\{ \frac{x^3}{2} - \frac{3x^2}{2} \right\} \cdot dx$$

$$= \left[\frac{1}{8}x^4 - \frac{1}{2}x^3 \right]_{3}^{5}$$

$$= \left(\frac{625}{8} - \frac{125}{2} \right) - \left(\frac{81}{8} - \frac{27}{2} \right) = 19$$

remember that $\mu = \text{E}(X) = \dfrac{13}{3}$

so $\sigma^2 = \text{E}(X^2) - \mu^2 = 19 - \left(\dfrac{13}{3} \right)^2 = \dfrac{2}{9}$

and the standard deviation of X is $\sqrt{\dfrac{2}{9}} = 0.471\,(3\text{s.f.})$.

The standard deviation is the square root of the variance.

Example 8

The continuous random variable X has probability density function

$\qquad f(x) = kx \quad$ for $0 \le x \le 5$

a) Find the value of k.

b) Find the mean and variance of X.

c) Calculate

   i) $\text{P}(X > \mu)$ and

   ii) $\text{P}(X > \mu + \sigma)$ where μ is the mean and σ is the standard deviation of X.

▶ Continued on the next page

a) $F(x) = \begin{cases} 0 & x < 0 \\ \dfrac{k}{2}x^2 & 0 \le x \le 5 \text{ so } \dfrac{k}{2} \times 5^2 = 1 \implies k = \dfrac{2}{25} \\ 1 & x > 5 \end{cases}$

> $F(x)$ is the *cumulative* probability function and is found by integrating the pdf. It is equal to zero below the lower limit of the pdf and 1 above the upper limit. In this case, $F(5) = 1$ so we can calculate k directly.

b) $\mu = E(X) = \displaystyle\int_{-\infty}^{\infty} x f(x) \cdot dx$

$$= \int_0^5 \left\{ x \cdot \frac{2}{25}x \right\} \cdot dx$$

$$= \int_0^5 \left\{ \frac{2}{25}x^2 \right\} \cdot dx$$

$$= \left[\frac{2}{75}x^3 \right]_0^5 = \frac{2 \times 125}{75} - 0 = \frac{10}{3}$$

$E(X^2) = \displaystyle\int_{-\infty}^{\infty} x^2 f(x) \cdot dx$

$$= \int_0^5 \left\{ x^2 \cdot \frac{2}{25}x \right\} \cdot dx$$

$$= \int_0^5 \left\{ \frac{2}{25}x^3 \right\} \cdot dx$$

$$= \left[\frac{1}{50}x^4 \right]_0^5 = \frac{625}{50} - 0 = 12.5$$

and $\sigma^2 = E(X^2) - \mu^2 = 12.5 - \left(\dfrac{10}{3} \right)^2 = \dfrac{25}{18}$

c) i) $P(X > \mu) = 1 - F\left(\dfrac{10}{3} \right) = 1 - \dfrac{1}{25}\left(\dfrac{10}{3} \right)^2 = \dfrac{5}{9}$

ii) $P(X > \mu + \sigma) = 1 - F\left(\dfrac{10}{3} + \sqrt{\left(\dfrac{25}{18} \right)} \right) = 1 - \dfrac{1}{25}\left(\dfrac{10}{3} + \sqrt{\left(\dfrac{25}{18} \right)} \right)^2 = 0.186 (3\text{s.f.})$

Advice on calculator use

Note: You can work out a decimal value for $\mu + \sigma$ but you need to keep enough decimal places in it to give answers to 3 s.f. The best way of doing this is that if you calculate $\mu + \sigma = 4.5118\dots$ and that is the last calculation done on your calculator, then the **exact** value can be used by pressing the **Ans** key.

Example 9

The continuous random variable X has probability density function

$$f(x) = \frac{1}{36}(9 - x^2) \qquad \text{for } -3 \leq x \leq 3$$

a) Find the mean and variance of X.

b) Calculate   i) $P(X > 2)$ and   ii) $P(|X| > \sigma)$ where σ is the standard deviation of X.

a) $\mu = E(X) = \displaystyle\int_{-\infty}^{\infty} xf(x)\cdot dx$

$$= \int_{-3}^{3} \left\{ x \cdot \frac{1}{36}(9 - x^2) \right\} \cdot dx$$

$$= \int_{-3}^{3} \left\{ \frac{1}{36}(9x - x^3) \right\} \cdot dx$$

$$= \left[\frac{1}{36} \left(\frac{9}{2}x^2 - \frac{x^4}{4} \right) \right]_{-3}^{3}$$

$$= \frac{1}{36} \left\{ \left(\frac{81}{2} - \frac{81}{4} \right) - \left(\frac{81}{2} - \frac{81}{4} \right) \right\} = 0$$

(Note that this could have been found by the symmetry of the pdf.)

$$E(X^2) = \int_{-\infty}^{\infty} x^2 f(x) \cdot dx$$

$$= \int_{-3}^{3} \left\{ x^2 \cdot \frac{1}{36}(9 - x^2) \right\} \cdot dx$$

$$= \int_{-3}^{3} \left\{ \frac{1}{36}(9x^2 - x^4) \right\} \cdot dx$$

$$= \left[\frac{1}{36} \left(3x^3 - \frac{x^5}{5} \right) \right]_{-3}^{3}$$

$$= \frac{1}{36} \left\{ \left(81 - \frac{243}{5} \right) - \left(-81 - \frac{-243}{5} \right) \right\}$$

$$= 1.8$$

and $\sigma^2 = E(X^2) - \mu^2 = 1.8(-0^2) = 1.8$

▶ Continued on the next page

b) i) $P(X > 2) = \int_2^\infty f(x) \cdot dx = \int_2^3 \left\{ \frac{1}{36}(9 - x^2) \right\} \cdot dx$

$$= \left[\frac{1}{4}x - \frac{1}{108}x^3 \right]_2^3$$

$$= \left(\frac{3}{4} - \frac{27}{108} \right) - \left(\frac{1}{2} - \frac{8}{108} \right) = \frac{2}{27}$$

ii) $P(|X| > \sigma)$ will be $2 \times P(X > \sigma)$ by symmetry, and $\sigma = \sqrt{1.8}$ from part **a**.

$$P(X > \sigma) = \int_{\sqrt{1.8}}^3 f(x) \cdot dx = \int_{\sqrt{1.8}}^3 \left\{ \frac{1}{36}(9 - x^2) \right\} \cdot dx$$

$$= \left[\frac{1}{4}x - \frac{1}{108}x^3 \right]_{\sqrt{1.8}}^3$$

$$= \left(\frac{3}{4} - \frac{27}{108} \right) - \left(\frac{\sqrt{1.8}}{4} - \frac{(\sqrt{1.8})^3}{108} \right) = 0.18695\ldots$$

and $P(|X| > \sigma) = 0.374$ (3 s.f.)

Example 10

The weekly petrol consumption, in hundreds of litres, of a sales representative
may be modelled by the random variable X with probability density function

$$f(x) = ax^2(b - x) \qquad \text{for } 0 \le x \le 2.$$

a) Find the values of a and b if the mean consumption is 144 litres.

b) Find the standard deviation of the weekly petrol consumption.

• •

a) The total probability is 1, so $1 = \int_{-\infty}^\infty f(x) \cdot dx$

$$1 = \int_0^2 ax^2(b - x)\, dx = \left[\frac{ab}{3}x^3 - \frac{a}{4}x^4 \right]_0^2$$

$$= \left(\frac{8ab}{3} - 4a \right) - 0$$

$$\mu = E(X) = \int_{-\infty}^\infty xf(x) \cdot dx$$

$$= \int_0^2 \{ x \cdot ax^2(b - x) \} \cdot dx$$

$$= \int_0^2 \{ abx^3 - ax^4 \} \cdot dx$$

$$= \left[\frac{ab}{4}x^4 - \frac{a}{5}x^5 \right]_0^2 = \left(4ab - \frac{32a}{5} \right) - 0 = 1.44$$

Solving the linear simultaneous equations in a and b gives $a = 0.15$, $b = 4$

b) $E(X^2) = \int_{-\infty}^{\infty} x^2 f(x) dx$

$= \int_{0}^{2} \left\{ x^2 \cdot \frac{3}{20} x^2 \right\} (4 - x) dx$

$= \int_{0}^{2} \left\{ \frac{3}{5} x^4 - \frac{3}{20} x^5 \right\} dx$

$\left[\frac{3}{25} x^5 - \frac{1}{40} x^6 \right]_{0}^{2} = \left(\frac{96}{25} - \frac{64}{40} \right) - 0 = \frac{56}{25}$

$\sigma^2 = \frac{56}{25} - 1.44^2 = 0.1664$

So the standard deviation of X is 0.408 and the standard deviation of the petrol consumption is 40.8 litres (3 s.f.).

Exercise 5.3

1. A probability density function is given by $f(x) = \frac{2x}{9}$ for $4 \leq x \leq 5$.

   Find the mean and variance of X.

2. The continuous random variable X has probability density function

   $f(x) = kx \quad$ for $0 \leq x \leq 3$

   a) Find the value of k.

   b) Find the mean and variance of X.

   c) Calculate

       i) $P(X > \mu)$ and

       ii) $P(X > \mu + \sigma)$ where μ is the mean and σ is the standard deviation of X.

3. The continuous random variable X has probability density function

   $f(x) = kx^3 \quad$ for $0 \leq x \leq 5$

   a) Find the value of k.

   b) Find the mean and variance of X.

   c) Calculate

       i) $P(X > \mu)$ and

       ii) $P(X > \mu - 2\sigma)$ where μ is the mean and σ is the standard deviation of X.

4. The continuous random variable X has probability density function

$f(x) = kx^4$ for $0 \leq x \leq 1$

a) Find the value of k.

b) Find the mean and variance of X.

c) Calculate

    i) $P(X < \mu)$ and

    ii) $P(|X - \mu| < \sigma)$ where μ is the mean and σ is the standard deviation of X.

5. The length, in metres, of jumps that Carl makes may be modelled by the probability density function

$$f(x) = k(x - 6)^2 \quad \text{for } 5.5 \leq x \leq 6.5$$

a) State the value of $E(X)$.

b) Find the value of k.

c) Find the variance of X.

6. Buses arrive regularly at a bus stop every 15 minutes.

a) Gupta does not know the bus schedule so he arrives at the bus stop at random times.

    i) If X is the length of time, in minutes, that Gupta has to wait for a bus, explain why $f(x) = \frac{1}{15}$ for $0 \leq x \leq 15$ is a sensible model for the distribution.

    ii) Find the mean and variance of X.

b) Shirma does know the bus schedule and aims to arrive shortly before a bus is due. If Y is the length of time, in minutes, that Shirma has to wait for a bus, and Y has probability density function given by

$f(y) = k(3 - y)$ for $0 \leq y \leq 3$

    i) Find the value of k.

    ii) Find the mean and variance of Y.

7. The profit, X, *in* $000s, made on a speculative investment may be modelled by the probability density function given by $f(x) = k(9 + 8x - x^2)$ for $-1 \leq x \leq 4$.

a) Show that $k = \frac{3}{250}$.

b) Find the mean and standard deviation of the profit made on the investment.

c) Show that the probability of making a loss on the investment is $\frac{6}{125}$.

d) There is an alternative investment on offer with a guaranteed return of $1500. Find the probability that the speculative investment gives a better return.

8. The diameters, in cm, of chips made in car paintwork by flying stones during resurfacing of a road, may be modelled by the random variable X with probability density function $f(x) = ax^2(b - x)$ for $0 \le x \le 3$.

   a) Show that $a = \dfrac{4}{27}$ and $b = 3$ if the mean diameter is 1.8 cm.

   b) Find the standard deviation of the diameters.

   c) Comment on whether you think the pdf given is likely to be a good model to describe the diameters.

5.4 Mode of a continuous random variable

The mode of a continuous random variable is the value of x for which $f(x)$ is a maximum over the interval in which $f(x)$ exists.

This is either a stationary point, at which $f'(x) = 0$, or the end value of the interval over which $f(x)$ is defined.

Note: This section is not within the syllabus but is included for completeness.

A sketch of the pdf will help you to make sense of the mode.

There may not be a mode if no single value occurs more often than any other (sometimes extended to talking about two modes but not normally any more).

> **Example 11**
> $f(x) = \dfrac{1}{39}(9 - x^2)$ for $-3 \le x \le 3$
> Find the mode.
> ....
> Look at a sketch of the pdf:
>
>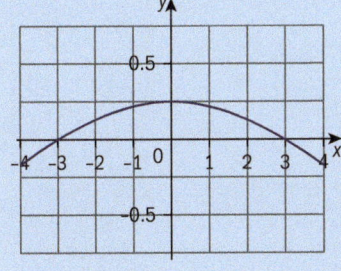
>
> The mode is at 0.
>
> You could alternatively use differentiation:
> $f(x) = \dfrac{1}{36}(9 - x^2) \Rightarrow f'(x) = \dfrac{-2x}{36} = 0$ when $x = 0$.

Example 12

Find the mode of the pdf from Example 1 in Section 5.2,

$$f(x) = \frac{1}{2}(x - 3) \quad \text{for } 3 \leq x \leq 5$$

Sketch the graph:

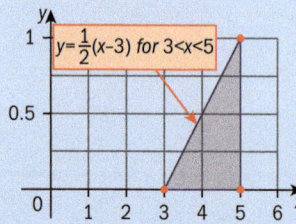

The derivative of $f(x)$ here is just $\frac{1}{2}$ which has no dependence on x so there will be no stationary value of $f'(x)$. This does not mean there is no mode.

The pdf increases steadily throughout the interval in which it is defined. The mode for the random variable is 5 (the upper bound of the interval in which $f(x)$ is defined).

Exercise 5.4

1. The continuous random variable X has probability density function given by
    $$f(x) = \frac{3}{2}(x - 1)^2 \quad 0 \leq x \leq 2.$$

    a) Sketch the probability density function.

    b) State the value of

        i)   the mode

        ii) The median of X.

2. For the random variables with the following probability density functions, find **i)** the median

        ii) the mode

    a) $f(x) = \frac{1}{12}x \quad 1 \leq x \leq 5$

    b) $f(x) = \frac{x}{2} \quad 0 \leq x \leq 2$

    c) $f(x) = \frac{3}{8}x^2 \quad 0 \leq x \leq 2$

3. The continuous random variable X has probability density function given by
    $$f(x) = kx(10 - x) \quad 0 \leq x \leq 7.$$

    a) Find the value of k.

    b) Find the mode X.

Summary exercise 5

1. A continuous random variable X has a probability density function of the form $f(x) = A(x^2 + 4)$, $0 < x < 1$.
   a) Calculate the value of A.
   b) Find $P(0.5 \leq x \leq 1)$ correct to 2 decimal places.
   c) Calculate $E(X)$ and $Var(X)$.

2. A continuous random variable X has a probability density function of the form $f(x) = \frac{x}{6} + C$, $0 \leq x \leq 3$.
   a) Calculate the value of C.
   b) Find $P(1 \leq x \leq 2)$ correct to 2 decimal places.
   c) Calculate $E(X)$ and $Var(X)$.

3. A continuous random variable X has a probability density function of the form $f(x) = 1 - \frac{x}{2}$, $0 \leq x \leq 2$.
   a) Find $P(1 \leq x \leq 2)$.
   b) Calculate $E(X)$ and $Var(X)$.
   c) Calculate $P(X \leq \mu)$ where $\mu = E(X)$.

4. A factory produces discs to operate electronic games machines. The slots in the electronic games machines are 3 cm long and the factory is requested to make discs whose diameters are no more than 2.9 cm and no less than 2.7 cm. The probability density function of X, the disc diameter is
   $$f(x) = \frac{3000}{32}(3.1 - x)(x - 2.7), \ 2.7 \leq x \leq 3.1.$$
   A disc is chosen at random. Find:
   i) the probability that the disc does meet the required specification;
   ii) the probability that the disc does not meet the required specification, but will still fit in the slot.

5. A continuous random variable X has probability density function given by
   $$f(x) = \begin{cases} 4(1-x)^3 & 0 \leq x \leq 1 \\ 0 & \text{otherwise} \end{cases}$$
   Find
   a) $P(X > 0.5)$
   b) the mean and variance of X.

6. The continuous random variable X has probability density function given by
   $$f(x) = \begin{cases} k \cos x & 0 \leq x \leq \frac{1}{4}\pi, \\ 0 & \text{otherwise}, \end{cases}$$
   where k is a constant.
   a) Show that $k = \sqrt{2}$. [2]
   b) Find $P(X > 0.4)$. [2]
   c) Find the upper quartile of X. [3]
   d) Find the probability that exactly 3 out of 5 random observations of X have values greater than the upper quartile. [2]

   Cambridge International A Level and AS Mathematics 9709 Paper 71 Q5 November 2009

7. The time in hours taken for clothes to dry can be modelled by the continuous random variable with probability density function given by
   $$f(t) = \begin{cases} k\sqrt{t} & 1 \leq t \leq 4, \\ 0 & \text{otherwise}, \end{cases}$$
   where k is a constant.
   a) Show that $k = \frac{3}{14}$. [3]
   b) Find the mean time taken for clothes to dry. [4]
   c) Find the median time taken for clothes to dry. [3]

d) Find the probability that the time taken for clothes to dry is between the mean time and the median time. [2]

Cambridge International A Level and AS Mathematics 9709 Paper 7 Q7 November 2008

8. A random variable X has probability density function given by

$$f(x) = \begin{cases} \dfrac{k}{x-1} & 3 \le x \le 5, \\ 0 & \text{otherwise,} \end{cases}$$

where k is a constant.

a) Show that $k = \dfrac{1}{\ln 2}$. [4]

b) Find a such that $P(X < a) = 0.75$. [4]

Cambridge International A Level and AS Mathematics 9709 Paper 71 Q5 November 2012

9. The time in minutes taken by people to read a certain booklet is modelled by the random variable T with probability density function given by

$$f(t) = \begin{cases} \dfrac{1}{2\sqrt{t}} & 4 \le t \le 9, \\ 0 & \text{otherwise,} \end{cases}$$

a) Find the time within which 90% of people finish reading the booklet. [3]

b) Find $E(T)$ and $Var(T)$. [6]

Cambridge International A Level and AS Mathematics 9709 Paper 71 Q6 May/June 2013

10.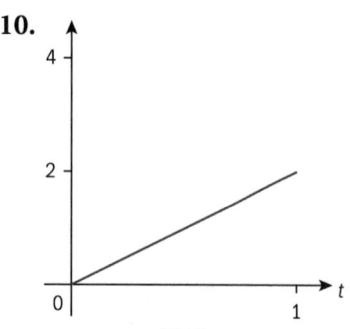

Fig. 1

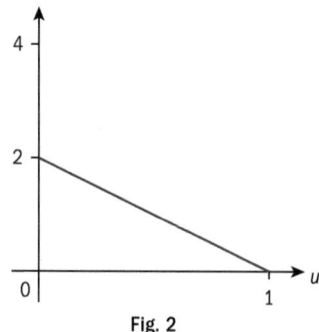

Fig. 2

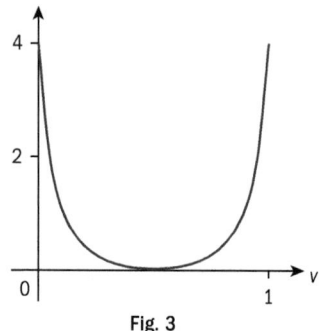

Fig. 3

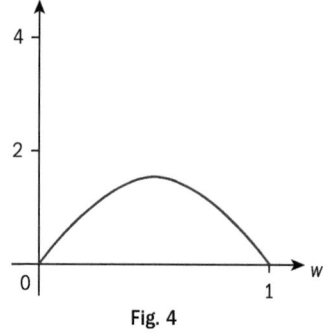

Fig. 4

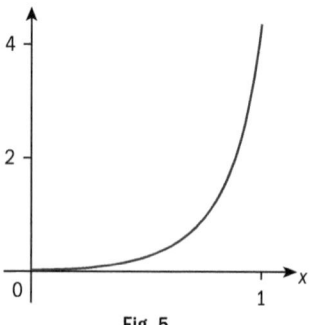

Fig. 5

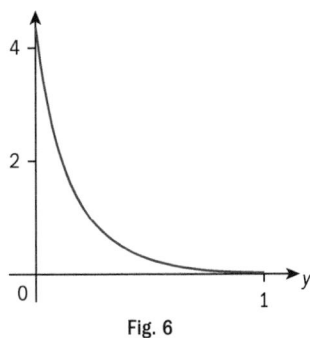

Fig. 6

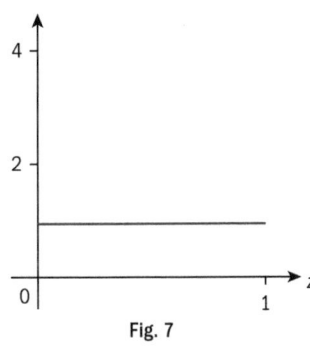

Fig. 7

Each of the random variables T, U, V, W, X, Y and Z takes values between 0 and 1 only. Their probability density functions are shown in Figs 1 to 7 respectively.

a) **i)** Which of these variables has the largest median? [1]

 ii) Which of these variables has the largest standard deviation? Explain your answer. [2]

b) Use Fig.2 to find $P(U < 0.5)$. [2]

c) The probability density function of X is given by

$$f(x) = \begin{cases} ax^n & 0 \le x \le 1, \\ 0 & \text{otherwise,} \end{cases}$$

where a and n are positive constants.

 i) Show that $a = n + 1$. [3]

 ii) Given that $E(X) = \dfrac{5}{6}$, find a and n. [4]

Cambridge International A Level and AS Mathematics 9709 Paper 71 Q7 November 2011

Chapter summary

- Continuous random variables are needed to describe situations involving length, time, mass, density, volume etc.

- A probability density function (pdf), $f(x)$ must satisfy these basic properties:

  $f(x) \ge 0$ for all x so that all probabilities are not negative

  $\displaystyle\int_{-\infty}^{\infty} f(x)\cdot dx = 1$ – often $f(x)$ is only defined over a small range, in which case the integral over that range will be 1.

- For a continuous random variable:

  $\mu = E(X) = \displaystyle\int_{-\infty}^{\infty} xf(x)\cdot dx$ where in practice the limits will be the limits of the interval over which $f(x)$ is defined.

  $Var(X) = \sigma^2 = E[\{X - E(X)\}^2]$ but you almost always use the computational form

  $\sigma^2 = E(X^2) - \mu^2$ where $\mu = E(X)$.

  $E(X^2) = \displaystyle\int_{-\infty}^{\infty} x^2 f(x)\cdot dx$

- The median of a continuous random variable X is α if $P(X \le \alpha) = 0.5$. This can be identified by geometrical arguments and can also be found by using integration.

Collecting data can be extremely costly, time consuming and even dangerous, so there is now a specialised area of research and development into sampling methods to provide cost-effective ways of producing the information required.

Applications range from the mundane, such as acceptance sampling techniques, which are used in industry when taking delivery from component suppliers, to cutting-edge science such as monitoring the effects of global warming.

Objectives

After studying this chapter you should be able to:

- understand the distinction between a sample and a population, and appreciate the necessity for randomness in choosing samples;
- explain in simple terms why a given sampling method may be unsatisfactory (knowledge of particular sampling methods, such as quota or stratified sampling, is not required, but candidates should have an elementary understanding of the use of random numbers in producing random samples);
- recognise that a sample mean can be regarded as a random variable, and use the facts that $E(\bar{X}) = \mu$ and that $Var(\bar{X}) = \dfrac{\sigma^2}{n}$;
- use the fact $\bar{X}$ has a normal distribution if X has a normal distribution;
- use the Central Limit theorem where appropriate;

Before you start

You should know how to:

1. List the possible outcomes in a simple experiment.
   I have a set of four cards with the numbers 1, 2, 3 and 4 on them. List the possible outcomes if I sample two cards at random without replacing the first
   1,2; 1,3; 1,4; 2,1; 2,3; 2,4; 3,1; 3,2; 3,4; 4,1; 4,2; 4,3

Skills check:

1. I have a set of four cards with the numbers 1, 2, 3 and 4 on them. List the possible outcomes if I sample two cards at random and replace the first before drawing the second card.

After studying this chapter you should be able to:

- understand the distinction between a sample and a population, and appreciate the necessity for randomness in choosing samples;

- explain in simple terms why a given sampling method may be unsatisfactory (knowledge of particular sampling methods, such as quota or stratified sampling, is not required, but candidates should have an elementary understanding of the use of random numbers in producing random samples);

- recognize that a sample mean can be regarded as a random variable, and use the facts that $E(\overline{X}) = \mu$ and that $\text{Var}(\overline{X}) = \dfrac{\sigma^2}{n}$;

- use the fact that $\overline{X}$ has a normal distribution if X has a normal distribution;

- use the Central Limit theorem where appropriate;

6.1 Populations, census and sampling

A sampling frame is a list containing all the **units or elements** which are the members of the **population** to be sampled. Because sampling is a practical activity, there will be times that the sampling frame will not match the population exactly,

e.g. a voting register may still include people who have died and may omit people who have completed the application forms but the forms are still being processed. A real life example occurred in 2007 when it was found that some TV companies were not including all the entrants who phoned into premium rate competition lines when drawing the competition winners.

The sample needs to be constructed so that **bias** is avoided, and ideally so that the sample will return an estimate very close to the true population parameter it is estimating.

A sample looks at some of the population.
A **census** examines the full population.

Random sampling

Each member of the population has an equal chance of being selected, and **all possible combinations are equally likely**.

Simple random sampling means sampling without replacement. In practice this is usually done by using random numbers or some other random process, e.g. taking names out of a hat – if it can be done in a genuinely random way.

You do not have to deal with different types of sampling methods in this course, but it is worth being aware that there are a lot of different methods developed to give the best possible estimates in different situations.

> Sections 6.1 to 6.3 are not commonly examined in depth but provide background understanding of key ideas in statistics. It is recommended that these sections are given a light touch.

Methods of collection

Data can be collected automatically by electronic measurement devices
e.g. in a hospital situation, or a production process, or manually through
surveys, questionnaires, and by direct observation and recording.

You need to take care when designing questionnaires, to ensure that questions are
unambiguous, and do not use language which will bias the respondent's answer.
They should be as simple to understand and answer as possible to collect the required
information, and you need to consider how the information is to be analysed to ensure
that the data is collected in a form which makes the analysis easily accessible.

Example 1

The population is all the students at an 11 – 18 school.
Suggest at least two different sampling frames which could be used for this population

....

Any system which lists all students in the school exactly once will be a
sampling frame:

- Alphabetical list of all students in school.
- Alphabetical lists of students by year group.
- Alphabetical lists of students by tutor group.
- List of students by date of birth, with alphabetical ordering for any
  students with the same date of birth.
- Any of these could take boys and girls separately and a different
  sampling frame would be produced.

Example 2

Suggest possible populations and sampling frames to investigate the following:
a) whether a university has a gender bias in its intake procedures.
b) where people in a town do their grocery shopping.

....

a) This is relatively straightforward. The population is anyone who applied to the university – for
   one or more years depending on how extensive the investigation will be. The corresponding
   sampling frame will be a list of these people (possibly alphabetical, but any list will do).

b) This is much less straightforward and highlights some of the practical problems faced in
   sampling. The population of interest is the people in the town who shop for groceries, but this
   is not very precisely defined – and there is no easily accessible way to produce a sampling
   frame which lists these people. Whatever registers (if any) which are held for voting, or tax
   collection, might give something which can be used as a 'best available' sampling frame – but
   these would vary from country to country. Any results obtained from surveys in a situation
   like this need to be treated with caution.

Exercise 6.1

1. **a)** Explain what you understand by the terms
       i) population
       ii) census.
    b) Give an example of bias in sampling.

2. **a)** Explain what you understand by the terms
       i) sampling frame
       ii) sampling unit.
    b) We want to collect data on age, ethnicity, education level and sex for adults in the UK.
       i) Describe the population, and a possible sampling frame in this situation.
       ii) Explain why the sampling frame is unlikely to exactly match the population.

3. Describe the important characteristics of a simple random sample.

4. A bank hires a market research company to conduct a survey of its customers to see how it could improve its service.
    a) State what the population is in this situation.
    b) Suggest a possible way to construct a sampling frame.
    c) List any difficulties you can identify in making sure that any individual customer does not appear more than once in the sampling frame.

6.2 Advantages and disadvantages of sampling

In some circumstances taking a census is simply not possible or desirable, e.g. in the run-up to elections, opinion polls are taken to provide some information on what people think on some issues. Here the population is everyone who could vote in the election and the amount of time (and money) required to ask everyone would be far too high – and the time required to process the information might mean that the election had already taken place before the analysis was available.

Moreover, even asking everyone would not guarantee a correct prediction because people are under no obligation to respond truthfully to this sort of opinion poll, or they may answer truthfully at the time and later change their mind and vote differently.

Sampling is generally much cheaper than taking a census, and almost always it is cost effective provided good sampling procedures are followed. A high proportion of the information available from a census can often be obtained from a sample which costs a fraction of what the census would cost.

In certain situations, testing items results in them being destroyed and a census is not appropriate. In many situations, it is important to have the analysis of the sample available quickly in order to inform decision making.

However, a sample will always give *incomplete information* while a census tells you everything about the population.

The key in sampling is to ensure the best possible quality of information for the resources that are available, and in particular to avoid sources of **bias**. There is **natural variation** between individual sampling units, which is due to chance, but there are often systematic differences between groups of sampling units with one or more shared characteristics, e.g. there are often differences between males and females.

Bias occurs where there is tendency for a sample selected to over or under represent certain groups – it is a property of the sampling method rather than the sample itself,
e.g. if a sample of 50 is taken by simple random sampling using a council tax register which has equal numbers of males and females on it, then even if the sample ends up with 22 men and 28 women there is no bias present. However, if the method used is to select a property at random and to take the first named person on the register at that property, then the sample will be biased because the first named person will more often be male. There are a number of common sources of bias:

- Subjective choice by person taking the sample

e.g. A person conducting a survey in a town centre may ask only people 'looking respectable' because they are thought more likely to agree to participate.

- Self-selection

e.g. Radio phone-in surveys are strongly biased. Only people who feel particularly strongly on the issue will participate.

- Non-response

Almost all surveys have an element of 'self-selection' in them because there is little the person conducting a survey can do to require people to take part. The *response rates* for postal surveys are critical in knowing how worthwhile any information will be when the analysis is done.

Companies will often offer inducements – entry into a prize draw for example – to encourage people to take part in their surveys.

- Convenience sampling (opportunity sampling)

Items are chosen by which is the easiest rather than in any structured and systematic attempt to construct an unbiased sample.

- Sampling from an incomplete sampling frame

e.g. If you use a telephone directory, or a list of registered voters as a sampling frame when the population is the adults in a community then there will be groups of people who are seriously under-represented.

You are not expected to know details of different sampling methods, but it may help to get a feel for what the issues are by reflecting on any good or bad things you can see in different methods. There is a section at the end of this chapter which lists very briefly the main features of a number of different sampling methods.

Exercise 6.2

1. An airline sends out a postal questionnaire to enquire about customers' views on a new route they are introducing. They say that all completed questionnaires will be entered into a draw to win free flights on this new route.
   a) Explain why the sample is biased.
   b) Explain why the airline might make this offer.

2. Give an example of a convenience sample.

3. A manufacturer of mobile phone batteries wants to know how long they will operate for before needing to be charged.
   a) Explain why a sample should be used rather than a census.
   The batteries are packed in boxes of 50 as they come off the production line. A manager suggests taking one of the boxes and sending them to be tested as the sample.
   b) Assuming 50 is a reasonable size of sample to take, do you think this is a good way to choose the batteries to be tested? Give any advantages and disadvantages you can see with the method and suggest an alternative if you do not think it is a good way to do it.

4. A bank hires a market research company to conduct a survey of its customers to see how it could improve its service.
   a) Explain why a sample is more appropriate here than a census.
   b) Can you think of any reasons why the bank might not want to just take a simple random sample of its customers?

6.3 Variability between samples and use of random numbers

If a number of people each asked a random sample of 4 pupils in your class some questions, how variable would the responses be? Does it depend on the question?

Think what answers you would get if you asked:

- How long did it take you to get to school this morning?
- How many DVDs do you own?
- Do you own a mobile phone?
- What is your favourite type of film at the moment?

It is easy to say that when you take a sample, natural variation means that repeating the sample will probably mean you get a different result, but having some feeling for how much difference there might be is important in knowing how to deal with the uncertainty that this introduces.

Here are the answers from a dozen pupils at one school.

Pupil identifier	Time to get to school (mins)	Number of DVDs	Own a mobile phone	Favourite type of film
A	2	43	Y	Action
B	15	25	Y	Comedy
C	10	21	Y	Horror
D	25	26	Y	Action
E	4	0	Y	Drama
F	10	36	Y	Comedy
G	5	89	Y	Horror
H	10	27	Y	Horror
I	15	34	Y	SciFi
J	15	29	Y	Action
K	20	31	Y	Drama
L	5	18	Y	Comedy

If you, and 5 friends, had each taken a sample of size 4 chosen at random from the pupils in this class (and this is a very small class really), would you all have got the same results? How much difference would there be? Is it the same for all the questions that were asked?

You could put the letters A–L on slips of paper in a hat, and each draw 4 out at random. There are other ways you can do the same thing, using random numbers – either from a table, or using a computer or calculator to generate them.

Because there are 12 to choose from, you need to use a two digit number, but there are actually 100 two digit numbers (00 to 99). The simplest way is to decide that getting 00, or anything from 13 to 99, you will ignore and take another number. If you get a number you have already chosen, you will pick another number until you have got 4 different numbers.

39	63	46	23	49	10	52	06
79	64	42	80	00	05	95	35
53	65	72	41	61	48	24	96
81	60	06	79	31	33	18	84
39	68	17	43	56	27	92	37
80	36	17	25	27	44	96	33
86	22	80	24	93	30	15	11
31	15	61	02	81	93	34	67

You can see how wasteful this is of random numbers in this case – it has taken until the 56th number for me to get a sample of size 4 – but if I was sampling from a group of 85 I would be ignoring many fewer values.

Starting at the top left and reading across, my sample numbers are 10, 6, 5 and 11 (6 occurred a second time, but was ignored). My sample then would be:

Pupil identifier	Time to get to school (mins)	Number of DVDs	Own a mobile phone	Favourite type of film
J(10)	15	29	Y	Action
F(6)	10	36	Y	Comedy
E(5)	4	0	Y	Drama
K(11)	20	31	Y	Drama

In my sample:

- the average time to get to school is 12 minutes (to the nearest minute)
- the average number of DVDs owned is 24
- 100% own a mobile phone
- there are 3 different film genres with Drama appearing twice.

A simple random sample is the equivalent of putting names in a hat and drawing them out blindly. All individuals in the population are equally likely to come out at any stage, and all possible combinations are equally likely.

Some simple arithmetic can allow you to be much more efficient in the use of random numbers. If you ignore only 00 and 97, 98, 99 in the 100 two digit numbers, and take the remainder when the number chosen is divided by 12, then all 96 numbers from 01 to 96 will map onto 1 to 12 with equal likelihood.

In this case you will get a different sample:

39	63	46	23	49	10	52	06
79	64	42	80	00	05	95	35
53	65	72	41	61	48	24	96

You can see how much less wasteful this is of the random numbers –you have only had to use 5 two digit numbers to get the sample of size 4.

39 gives a remainder of 3 – pupil C, so does the next number in the list (63), so it is passed over, then 46 gives a remainder of 10 – pupil J, 23 gives a remainder of 11 – pupil K, and 49 gives a remainder of 1 – pupil A.

Pupil identifier	Time to get to school (mins)	Number of DVDs	Own a mobile phone	Favourite type of film
C(3)	10	21	Y	Horror
J(10)	15	29	Y	Action
K(11)	20	31	Y	Drama
A(1)	2	43	Y	Action

In this sample:

- two of the pupils are the same and two are different
- the average time to get to school is again 12 minutes (to the nearest minute)
- the average number of DVDs owned is now 31
- 100% own a mobile phone
- there are 3 different film genres with Action appearing twice this time – but not the same three: Action appears and Comedy does not.

Tables of random digits are often presented in blocks of 5 to enable you to use them flexibly – for example if you want two digit numbers it is probably quicker to take the first pair and the last pair out of each block of five digits, than to be using all the digits. As long as you decide the rule by which you will identify the random numbers you will use then you can do whatever you find easy.

41089	57286	78925	86189	48509	86176
45087	14740	92741	10088	41571	97806
28446	39969	32627	54204	32209	59745

 Advice on calculator use
Random function on calculator (actually pseudo-random numbers): most calculators have at least one generator on them - the first one included is between 0 and 1, often to 3 decimal places, sometimes to rather more. These can be used to generate random integers between a and b by × (b − a + 1) and add a, taking the integer part of the answer e.g. for a number between 1 and 25 you multiply by 25, add 1, and take the integer part e.g. 0.341 → 8.525 → 9.525 → 9. On a programmable calculator a series of such random integers can be produced by **Int (Rnd# × 25 + 1)** and EXE repeatedly generates the series required.
You should make sure that you know what functions are on your calculator and how to use them.

Exercise 6.3

1. A Head of Year wants to find out where pupils would like to go on an end of year trip. He decides to ask a sample of 10 pupils out of the year group which has 90 pupils in it. Which of the following will give him a simple random sample?

   a) Put all the pupil names in a hat and draw out 10 names.

   b) List pupils alphabetically, choose a number from 1 to 9 at random to give the first pupil, and then take every 9th pupil in the list from there.

   c) Take the first 10 pupils who arrive at the next year assembly.

   d) List the pupils in order of their total examination score and take the first 10 pupils on the list.

   e) List the pupils in order of their total examination score and assign each a number from 1 to 90. Use a set of 2 digit random numbers to choose 10 pupils from the list (ignore 00 and 91–99 if they appear, and any repeats, and continue until you have 10 different numbers between 01 and 90 inclusive).

2. Use the table of random numbers below to generate simple random samples for each of the following situations.

   a) A sample of size 10 from a population of 90, starting at the top left and reading across.

   b) A sample of size 10 from a population of 90, starting at the top left and reading down.

c) A sample of size 5 from a population of size 42, starting at the top left and reading across.

d) A sample of size 6 from a population of size 38, starting on the left of the 3rd row and reading across.

52	98	44	17	71	20	17	63	47	88	22	02	18	9	93
54	73	99	99	17	50	60	78	94	55	73	39	58	00	98
07	67	88	88	97	76	20	65	22	97	16	96	60	11	65
08	64	14	80	32	76	25	84	30	92	59	01	65	49	68
43	71	56	2	82	26	27	16	95	87	77	03	75	35	18
49	46	58	74	98	26	73	80	79	19	50	87	35	43	63
24	35	79	43	81	42	34	24	01	41	10	63	74	97	65

3. Your calculator probably has a random number generator *Ran#* which generates a 3 digit decimal. Use this to take another 5 samples (each of size 4) from the class data in the section above, and generate similar summaries of the results to the one listed above.

Is there any difference in the variability of the answers to the 4 questions asked in the survey?

> **Advice on calculator use**
> Many calculators will display 0.230 only as 0.23 and 0.600 as 0.6, so if it shows less than 3 decimal places add one or two 0s.

4. a) Using the random number generator on your calculator, take a simple random sample of size 4 from the table below, and calculate the mean of the values in your sample.

Person	1	2	3	4	5	6	7	8	9	10	11	12	13	14	15	16	17	18	19	20
value	38	39	40	41	42	53	54	55	56	57	68	69	70	71	72	83	84	85	86	87

b) Take 4 more simple random samples from the same data set and calculate the mean of each (you will use these again in the next exercise so don't lose them).

5. Using the random number generator on your calculator, take a simple random sample of size 10 from the table below, and calculate the mean of your sample.

40	51	53	47	39	38	33	35	46	30	45	44	36	33	47
53	38	31	45	34	38	49	51	53	45	47	53	41	40	31
42	52	45	38	36	52	38	51	50	31	32	31	49	54	34
50	40	39	50	32	43	52	30	36	44	42	44	45	50	54
54	54	41	39	40	46	41	45	43	48	44	37	42	51	53
49	49	44	37	30	54	42	33	46	44					

There are 85 values, with 15 per row in the first 5 rows.
If you are in a class, look at the different estimates that other people have from their samples – the variability in these estimates.

6.4 The sampling distribution of a statistic

In this unit you can assume that, when it comes to analysing results of sampling, a simple random sampling method has been used. This ensures that all possible combinations of sampling units are equally likely to be chosen in the sample, the method is unbiased, and that the following ideas are valid.

If you throw a pair of dice repeatedly and take the average score on the two dice, you can generate the probability distribution for this quite easily. The probability distribution of X is:

See chapter 3 for the distribution of the sum of the scores on the two dice

x	1	1.5	2	2.5	3	3.5	4	4.5	5	5.5	6
$P(X = x)$	$\frac{1}{36}$	$\frac{2}{36}$	$\frac{3}{36}$	$\frac{4}{36}$	$\frac{5}{36}$	$\frac{6}{36}$	$\frac{5}{36}$	$\frac{4}{36}$	$\frac{3}{36}$	$\frac{2}{36}$	$\frac{1}{36}$

Think of sampling throws of a fair die.
If you throw the die n times and take the average of the scores, you have taken a simple random sample and for any value of n (in theory) you could work out a probability distribution.
The mean of the n throws is called a **statistic** (it has to depend only on the values observed in the random sample), and the probability distribution is known as the **sampling distribution of the statistic**.

Examples of other statistics are the median, the standard deviation of the sample, the range etc., and the sampling distributions for some of these can be surprisingly complicated. Fortunately the ones which are most useful have relatively simple distributions, or can be approximated by something simple.

Example 3

The heights of 15 year old males in a large town have a mean μ and a standard deviation σ. The heights of a sample of twenty 15 year old males and recorded as $x_1, x_2, x_3, \ldots x_{20}$.
Which of the following are statistics – for any which are not, explain why not.

a) $\displaystyle\sum_{i=1}^{20} x_i$ b) $\displaystyle\sum_{i=1}^{20} (x_i - \mu)^2$

c) $\displaystyle\sum_{i=1}^{20} (x_i - \bar{x})^2$ where $\bar{x} = \displaystyle\sum_{i=1}^{20} x_i \div 20$

d) the largest of the values $x_1, x_2, x_3, \ldots x_{20}$

e) the range of the sample values   f) $\dfrac{x_1 - \mu}{\sigma}$

g) the number of values $x_1, x_2, x_3, \ldots x_{20}$ which are greater than μ.

h) the number of values $x_1, x_2, x_3, \ldots x_{20}$ which are greater than $\bar{x}$.

....

A statistic must depend on some or all of the values of sample observations, and not on anything else, so **a**, **c**, **d**, **e** and **h** are all statistics.
b and **g** need the value of μ, and **f** needs the values of μ and σ, so these three are not statistics.

One of the simplest situations is when the 'measurement' of each member of a sample can be modelled by a random variable taking the value 1 when some criterion is satisfied and 0 when it is not.
E.g. If a person visiting a health centre is a male, or if a light bulb on a production line is faulty, or a bank manager intends to retire before their 60th birthday.
In this case, the sampling distribution of a sample of size n will be a Binomial random variable with parameters n and p where p is the probability the criterion is satisfied for any individual.

Example 4

20% of the trainee teachers in a college are male. A random sample of 15 teachers is taken from the college. The random variables $X_i: i = 1, 2, 3, \ldots. 15$ are defined as
$$X_i = \begin{cases} 1 \text{ if the } i\text{th trainee teacher is male} \\ 0 \text{ if the } i\text{th trainee teacher is female} \end{cases}$$

a) Write down the distribution for $\displaystyle\sum_{i=1}^{15} X_i$

b) Find $P\left(\displaystyle\sum_{i=1}^{15} X_i = 3\right)$

c) Give the values of $E\left(\displaystyle\sum_{i=1}^{15} X_i\right)$ and $Var\left(\displaystyle\sum_{i=1}^{15} X_i\right)$.

....

a) This is a Binomial with $n = 15$ and $p = 0.2$.

b) 0.250

c) $E\left(\sum_{i=1}^{15} X_i\right) = np = 15 \times 0.2 = 3; \ \mathrm{Var}\left(\sum_{i=1}^{15} X_i\right) = npq = 15 \times 0.2 \times 0.8 = 2.4$

In other situations, you may be able to derive a sampling distribution directly from listing possible outcomes and their associated probabilities.

Example 5

A bag contains three 50 cent coins, two 20 cent coins and a 10 cent coin. Two coins are to be taken from it. Find the sampling distribution of the mean value of the coins taken.

Construct a table showing the 30 (equally likely) possible outcomes of taking two coins from this bag, and what the mean value is, to give the sampling distribution:

X	50	50	50	20	20	10
50		50	50	35	35	30
50	50		50	35	35	30
50	50	50		35	35	30
20	35	35	35		20	15
20	35	35	35	20		15
10	30	30	30	15	15	

X	50	35	30	20	15
P(X = x)	$\dfrac{6}{30}$	$\dfrac{12}{30}$	$\dfrac{6}{30}$	$\dfrac{2}{30}$	$\dfrac{4}{30}$

From this very simple situation, you get a quite strange looking distribution

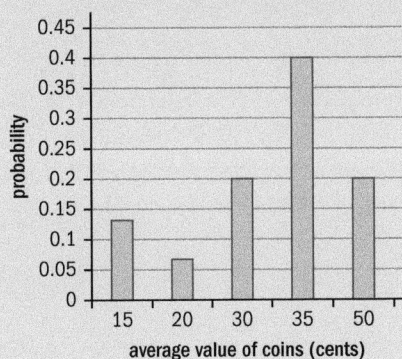

sampling distribution choosing two coins

Example 6

How would the sampling distribution be different if it was a really large bag of coins which had the 50p, £1 and £2 coins in the ratio of 3:2:1?

You could model this by thinking of using the same bag, but replacing the coin after noting the value.

The table and probability distribution then looks like this:

The change in likelihood has changed so little when there are a very large number of each type of coin.

X	50	50	50	20	20	10
50	50	50	50	35	35	30
50	50	50	50	35	35	30
50	50	50	50	35	35	30
20	35	35	35	20	20	15
20	35	35	35	20	20	15
10	30	30	30	15	15	10

X	50	35	30	20	15	10
$P(X = x)$	$\dfrac{9}{36}$	$\dfrac{12}{36}$	$\dfrac{6}{36}$	$\dfrac{4}{36}$	$\dfrac{4}{36}$	$\dfrac{1}{36}$

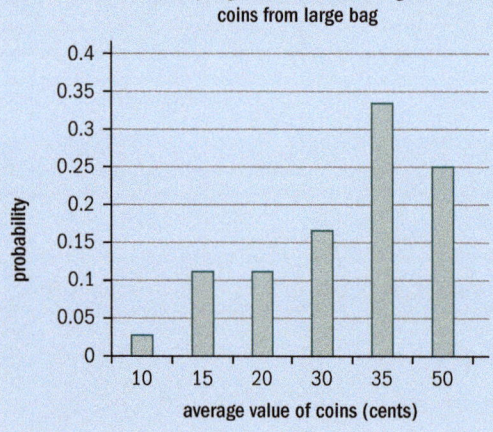

sampling distribution choosing two coins from large bag

Exercise 6.4

1. The incomes of doctors working for a large health organisation have a mean μ and a standard deviation σ. The incomes of a sample of twenty doctors working for the trust are recorded as $x_1, x_2, x_3, \dots x_{20}$.

   Which of the following are statistics? For any which are not, explain why not.

   a) $x_1 - x_2$

   b) $\sum_{i=1}^{20} (x_i - 80,000)^2$

   c) $\sum_{i=1}^{20} (x_i - \bar{x})^2$ where $\bar{x} = \sum_{i=1}^{20} x_i \div 20$

   d) the median of the values $x_1, x_2, x_3, \dots x_{20}$

   e) $\dfrac{\bar{x} - \mu}{\sigma}$

2. One third of the members of a fitness club are over 40 years old.
   A random sample of 20 members is taken from the club.
   The random variables X_i: $i = 1, 2, 3, \dots 20$ are defined as

   $$X_i = \begin{cases} 1 \text{ if the } i\text{th member is over 40} \\ 0 \text{ if the } i\text{th member is not over 40} \end{cases}$$

   a) Write down the distribution for $\sum_{i=1}^{20} X_i$

   b) Find $P\left(\sum_{i=1}^{20} X_i = 2\right)$

   c) Give the values of $E\left(\sum_{i=1}^{20} X_i\right)$ and $Var\left(\sum_{i=1}^{20} X_i\right)$.

3. Three quarters of the members of a cycle club own more than one bike.
   A random sample of 10 members is taken from the club.
   The random variables X_i: $i = 1, 2, 3, \dots 10$ are defined as

   $$X_i = \begin{cases} 1 \text{ if the } i\text{th member owns more than one bike} \\ 0 \text{ if the } i\text{th member does not own more than one bike} \end{cases}$$

   a) Write down the distribution for $\sum_{i=1}^{10} X_i$

   b) Find $P\left(\sum_{i=1}^{10} X_i \le 7\right)$

   c) Give the values of $E\left(\sum_{i=1}^{10} X_i\right)$ and $Var\left(\sum_{i=1}^{10} X_i\right)$.

4. A video game machine takes tokens for 1 game and for 3 games.
   20% of the tokens used are for 1 game.
   a) Find the mean μ for the number of games per token.

   A random sample of 3 tokens is taken when the machine is emptied.
   b) List all possible samples.

   c) Find the sampling distribution for the mean number of games per token.

   d) Find the sampling distribution for the mode of the number of games per token.

   e) Find the sampling distribution for the median of the number of games per token.

5. Another video game machine takes tokens for 1 game, for 3 games and for 6 games.

   The number of tokens used are in the ratio 1:2:2 for 1, 3 and 6 games.
   a) Find the mean μ for the number of games per token.
   A random sample of 2 tokens is taken when the machine is emptied.

   b) List all possible samples and the probability that it is taken.

   c) Find the sampling distribution for the mean number of games per token.

   d) Find the sampling distribution for the median of the number of games per token.

6.5 Sampling distribution of the mean of repeated observations of a random variable

In chapter 3 you looked at linear combinations of random variables and had the following results:

1: $E(aX + b) = aE(X) + b$
2: $Var(aX + b) = a^2 Var(X)$
3: $E(aX + bY) = aE(X) + bE(Y)$
4: $Var(aX + bY) = a^2 Var(X) + b^2 Var(Y)$ [if X, Y are independent]

If n independent observations are taken of a random variable X which has

mean μ and variance σ^2, and $\bar{X} = \dfrac{1}{n}\left(\sum_{i=1}^{n} X_i\right)$ is the sample mean then $\bar{X}$ is a

random variable and we can apply the results above to deduce that

$$E(\bar{X}) = \mu; \quad Var(\bar{X}) = \frac{\sigma^2}{n}$$

$\sum_{i=1}^{n} X_i = X_1 + X_2 + \dots + X_n$ so applying 3 above gives

$$E\left(\sum_{i=1}^{n} X_i\right) = E(X_1 + X_2 + \dots + X_n) = \mu + \mu + \dots + \mu = n\mu$$

And applying 4 above gives

$$\text{Var}\left(\sum_{i=1}^{n} X_i\right) = \text{Var}(X_1 + X_2 + \dots + X_n) = \sigma^2 + \sigma^2 + \dots + \sigma^2 = n\sigma^2$$

Then we can apply 1 and 2 above to go from the expectation and variance of the sum to the expectation and variance of the sample mean.

$$E(\bar{X}) = E\left(\frac{1}{n}\sum_{i=1}^{n} X_i\right) = \frac{1}{n}E\left(\sum_{i=1}^{n} X_i\right) = \frac{n\mu}{n} = \mu$$

$$\text{Var}(\bar{X}) = \text{Var}\left(\frac{1}{n}\sum_{i=1}^{n} X_i\right) = \frac{1}{n^2}\text{Var}\left(\sum_{i=1}^{n} X_i\right) = \frac{n\sigma^2}{n^2} = \frac{\sigma^2}{n}$$

Note that here it is helpful to talk about the expected value (or expectation) of the sample mean, where in most cases you would refer to the mean of the random variable rather than its expected value.

Example 7

i)   Find the expectation and variance of the score showing on a fair die.

ii) Find the probability distribution for the mean score when two fair dice are thrown.

iii) Find the expectation and variance of the mean score showing on two fair dice.

iv) Show that the expectations in **i)** and **iii)** are equal and the variance in **iii)** is half of that in **i)**

....

i) $E(X) = \frac{1}{6}(1 + 2 + 3 + 4 + 5 + 6) = 3.5$

$E(X^2) = \frac{1}{6}(1 + 4 + 9 + 16 + 25 + 36) = \frac{91}{6} \Rightarrow \text{Var}(X) = \frac{91}{6} - 3.5^2 = \frac{35}{12}$

ii) From the start of section 6.4 the distribution is:

$\bar{x}$	1	1.5	2	2.5	3	3.5	4	4.5	5	5.5	6
$P(\bar{X} = \bar{x})$	$\frac{1}{36}$	$\frac{2}{36}$	$\frac{3}{36}$	$\frac{4}{36}$	$\frac{5}{36}$	$\frac{6}{36}$	$\frac{5}{36}$	$\frac{4}{36}$	$\frac{3}{36}$	$\frac{2}{36}$	$\frac{1}{36}$

iii) $E(X) = \frac{1}{36}(1 + 3 + 6 + 10 + 15 + 21 + 20 + 18 + 15 + 11 + 6) = 3.5$

$E(X^2) = \frac{1}{36}(1 + 4.5 + 12 + 25 + 45 + 73.5 + 80 + 81 + 75 + 60.5 + 36) = \frac{329}{24}$

$\Rightarrow \text{Var}(X) = \frac{329}{24} - 3.5^2 = \frac{35}{24}$

iv) $E(X) = 3.5$ in both cases and $\frac{35}{24} = \frac{1}{2} \times \frac{35}{12}$ as required.

Example 8

A random variable has probability distribution given by

x	1	2	3	4
p	$5k$	$2k$	k	$2k$

i) Show that $k = 0.1$ and calculate $E(X)$ and $Var(X)$

ii) If $\bar{X}$ is the mean of a randomly selected sample of 5 observations of x, write down the expectation and variance of $\bar{X}$.

i) $5k + 2k + 2k = 10k = 1 \implies k = 0.1,$

x	1	2	3	4
p	0.5	0.2	0.1	0.2

$E(X) = 1 \times 0.5 + 2 \times 0.2 + 3 \times 0.2 + 4 \times 0.2 = 2.3$

$E(X^2) = 1 \times 0.5 + 4 \times 0.2 + 9 \times 0.2 + 16 \times 0.2 = 6.3 \implies Var(X) = 6.3 - 2.3^2 = 1.01$

ii) $E(\bar{X}) = E(X) = 2.3;\ Var(\bar{X}) = \frac{1}{5}Var(X) = \frac{1.01}{5} = 0.202$

Exercise 6.5

1. For the following random variables, work out the mean and variance of the mean of a sample of 10 independent observations of the random variable,

   a) $E(X) = 5;\ Var(X) = 4$

   b) $E(X) = 26.3;\ Var(X) = 7.8$

   c) Z is a random variable with probability distribution given by
   $Pr\{Z = z\} = \frac{5-z}{10};\ z = 1, 2, 3, 4.$

   d) X is a continuous random variable with pdf given by
   $f(x) = \frac{1}{36}(9 - x^2)$ for $-3 \le x \le 3$

2. X is a random variable with mean 42.5 and variance 23.1.

   a) Find the variance of the mean of a sample of 15 independent observations of X.

   b) What is the minimum size of sample needed in order that the variance of the sample mean is less than 1?

3. A fair spinner is equally likely to land on any of the four sections. The sections score 1, 4, 6 and 9 respectively.

   a) Find the mean and variance of the score obtained on a single spin.

   A random sample of 12 spins is taken and $\bar{X}_{12}$ denotes the mean of the twelve scores obtained.

   b) Find the mean and variance of $\bar{X}_{12}$

6.6 Sampling distribution of the mean of a sample from a Normal distribution

In section 4.2, you saw that the sum of two Normal distributions was still a Normal distribution. It follows that the distribution of $\bar{X} = \frac{1}{n}\sum_{i=1}^{n} X_i$ when $X \sim N(\mu, \sigma^2)$ will be $\bar{X} \sim N\left(\mu, \frac{\sigma^2}{n}\right)$ - that is, still Normal, with the same mean as the underlying population but the variance will be reduced by a factor of $\frac{1}{n}$.

Example 9

Packets of cereal are labelled as containing 500 grams of cereal. The actual contents can be modelled by a Normal distribution with mean 503 g and standard deviation 7 g.

What is the probability that the mean contents of a randomly selected sample of 10 packets will be under 498 g?

$$X \sim N\left(503, 7^2\right) \Rightarrow \bar{X} \sim N\left(503, \frac{7^2}{10}\right)$$

The mean stays at 503, but the variance is reduced by a factor of 10 because the sample size is 10.

$$P\left(\bar{X} < 498\right) = \Phi\left(\frac{498 - 503}{\frac{7}{\sqrt{10}}}\right) = \Phi(-2.259)$$

This is just the standard Normal distribution probability calculation from S1, chapter 8.

$$= 1 - \Phi(2.259) = 1 - 0.9881 = 0.0119 = 1.2\%$$

Example 10

The volume of liquid dispensed into bottles is normally distributed with mean 500 ml and standard deviation 6.2 ml.

i)   A random sample of 5 bottles is taken. What is the probability that the mean volume in the 5 bottles is at least 504 ml?

ii) Another random sample is to be taken. What is the minimum sample size needed for there to be no more than a 1% probability that the mean volume is greater than 501 ml?

▶ Continued on the next page

i) $X \sim N(500, 6.2^2) \Rightarrow \bar{X} \sim N\left(500, \dfrac{6.2^2}{10}\right)$ ← The mean stays at 500, but the variance is reduced by a factor of 5 because the sample size is 5.

$P(\bar{X} > 504) = 1 - \Phi\left(\dfrac{504 - 500}{\dfrac{6.2}{\sqrt{5}}}\right) = 1 - \Phi(1.443)$

$= 1 - \Phi(1.443) = 1 - 0.9255 = 0.0745 = 7.5\%$

ii) $X \sim N(500, 6.2^2) \Rightarrow \bar{X}_n \sim N\left(500, \dfrac{6.2^2}{n}\right)$

$P(\bar{X}_n > 501) < 0.01 \Rightarrow \Phi\left(\dfrac{501 - 500}{\dfrac{6.2}{\sqrt{n}}}\right) = 0.99$

$\Rightarrow \dfrac{501 - 500}{\dfrac{6.2}{\sqrt{n}}} = 2.326$ ← 2.326 is the z score cutting off the top 1% of a Normal distribution. Don't forget that the standard error has the factor of $\sqrt{n}$ and to find n you need to square – and then take the integer above the solution

$\Rightarrow \sqrt{n} = 2.326 \times 6.2 = 14.42\ldots \Rightarrow n = 207.91\ldots$

So a sample of size 208 is needed for there to be no more than a 1% probability that the mean volume is greater than 501 ml.

Exercise 6.6

1. The contents of bottles of water are normally distributed with mean 600 millilitres and standard deviation 7.2 millilitres.

   a) Give the distribution of the mean content of a random sample of 6 bottles.

   b) Find the probability that the mean content of a random sample of 6 bottles is less than 597 millilitres.

2. Packets of biscuits are labelled as containing 350 grams. The actual contents can be modelled by a Normal distribution with mean 352 grams and standard deviation 4.5 grams.

   a) Find the probability that the mean contents of a random sample of 10 packets is at least 350 grams.

   b) What is the smallest size of random sample for which there is probability of under 1% that the mean contents is under 350 grams?

3. $X \sim N\left(\mu, \sigma^2\right)$. $\bar{X}_n$ is the mean of a random sample of n independent observations of X. Given that $P\left(\left|\bar{X}_n - \mu\right| > 0.25\sigma\right) < 0.1$

   a) Find the smallest possible value of n

   b) For this value of n find $P\left(\left|\bar{X}_n - \mu\right| < 0.1\sigma\right)$

4. The lifetime of Sooperstrong batteries is normally distributed with mean 85 hours and standard deviation 9.2 hours.

   a) Find the probability that the mean lifetime of a random sample of 25 Sooperstrong batteries is less than 83 hours.

   The lifetime of Powersure batteries is normally distributed with mean 83 hours and standard deviation 2.1 hours.

   b) Find the probability that the mean lifetime of a random sample of 25 Sooperstrong batteries is shorter than the mean lifetime of a random sample of 5 Powersure batteries.

5. The number of characters on a page of a book can be modelled by a normal distribution with mean 1983 and standard deviation 44.7. Find the probability that the average number of characters per page in a random sample of 8 pages is more than 2000.

Compare this question with Q7 in the summary exercise of chapter 4.

6.7 The Central Limit Theorem

In section 6.5 you saw that the expectation and variance of a sample of n independent observations of a random variable X are known exactly from the expectation and variance of the random variable X. In section 4.1 you saw that the distribution of the sum of two independent Poisson random variables is still a Poisson, and in section 4.2 that the sum of two Normal distributions is another Normal, but in general the distribution of the sum of repeated observations of a random variable is not the same distribution as the original population. You have seen examples of this with the total score on two dice: for one die the distribution is uniform – all observations are equally likely, but for two dice the distribution is triangular shaped – starting at 2 it increases to a peak at the middle value of 7 and then decreases down to 12.

This means that while you know what the expectation and variance will be of the sample mean generally, you cannot make any probabilistic statements in the way that you could in examples 9 and 10.

The Central Limit Theorem says that the distribution of sample means becomes approximately Normal as the sample size increases, no matter what the underlying population is, and at this level an arbitrary cutoff of $n = 30$ is applied as a standard rule of thumb as to what sample size is needed in order to use the Central Limit Theorem.

It is worth exploring this a little further however, to get a firmer understanding of the theorem. For a uniform distribution, or for a symmetric binomial distribution [and for almost any reasonably symmetric population] the distribution of sample means is approximately Normal with much smaller sample sizes than 30.

There are two important facts which get lost sight of very often:

- The CLT says nothing about the standard error of the mean – that result is derived from algebraic manipulations of the definitions of mean and variance – the CLT is solely about the probability distribution of the sample mean.

- For any underlying population, increasing the sample size will make the sampling distribution of the mean look more like the Normal. No matter how strange the underlying distribution is, by the time the sample size gets to 30 the distribution will be passably close to the Normal distribution.

Except when the underlying population is itself Normal, in which case the sampling distribution of the mean will also be an exact Normal

Let's look at some instances:

Distribution of Sample Means

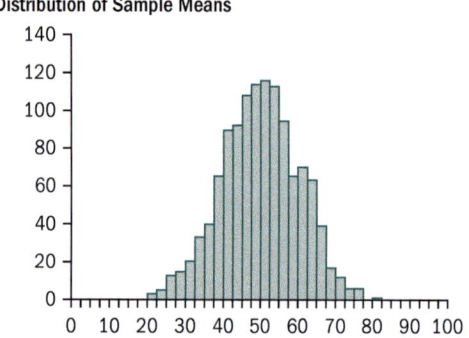

Samples = 1199   Mean = 50.0   St.dev. = 10.2

Samples of size 8 from Uniform distribution

This is a screenshot of a simulation after almost 1200 samples of size 8 had been taken from a Uniform distribution. The histogram of the sample mean values shows the characteristics of the Normal bell shaped curve.

What about a non-symmetric distribution? The graph below shows the pdf of the exponential distribution with a mean of 20.

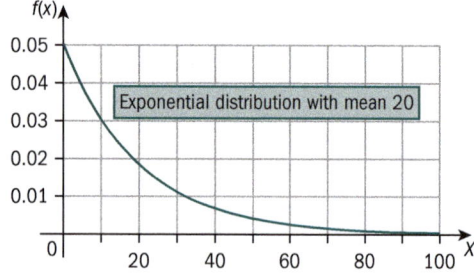

Exponential distribution with mean 20

And the screen shot below shows a simulation after almost 1300 samples of size 8 had been taken from this exponential distribution.

Distribution of Sample Means

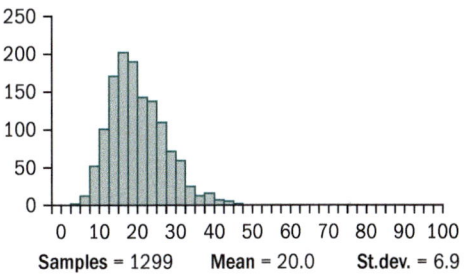

Samples = 1299   Mean = 20.0   St.dev. = 6.9

Samples of size 8 from the exponential distribution

For a sample size of only 8, it has moved a long way towards the Normal but not far enough yet that you would say that the Normal is a good approximation – it still has a noticeable positive skew.

By the time the sample size is 20, the skewness has largely disappeared from the sampling distribution, and using a Normal distribution would seem reasonable.

Distribution of Sample Means

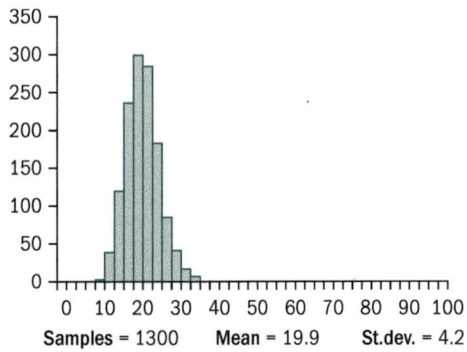

Samples = 1300   Mean = 19.9   St.dev. = 4.2

Samples of size 20 from the exponential distribution

Example 11

On a flight there are 43 bags checked in. Historical records show that on that route the mean weight of checked bags is 12.1 kg with a standard deviation of 3.8 kg.

What is the probability that the mean contents of the checked bags on this flight will be under 11 kg? (You may treat the bags on this flight as though they are a random sample of bags checked in on that route.)

Random sample of size 43 means $\bar{X} \underset{approx}{\sim} N\left(12.1, \frac{3.8^2}{43}\right)$

Even though you know nothing about the distribution of the weights of bags themselves, the CLT allows you to use this approximation.

$P(\bar{X} < 11) = \Phi\left(\dfrac{11 - 12.1}{\frac{3.8}{\sqrt{43}}}\right) = \Phi(-1.898)$

$= 1 - \Phi(1.898) = 1 - 0.9712 = 0.0288 = 2.9\%$

Exercise 6.7

1. The times taken in the cycling phase of a triathlon have mean 43.2 minutes and standard deviation 4.3 minutes. Find the probability that the mean cycling time taken by a random sample of 50 competitors is over 44 minutes.

2. The mean score in the 100 metres discipline of the decathlon is 912 points and the standard deviation is 42 points. Find the probability that the mean score of 45 randomly chosen decathletes in the 100 metres is over 900 points.

3. The times for a particular journey during rush hour in a large city have mean 23.3 minutes and standard deviation 8.9 minutes.

   a) Find the probability that the mean time taken for a random sample of 60 of the times for this journey is under 25 minutes.

   b) If you wanted to know the probability that the mean time taken for a random sample of 10 of the times for this journey was under 25 minutes, explain why you would not be able to give a reliable estimate.

4. $X \sim \text{Po}(7)$. A random sample of 72 observations of X is taken.

   i) State the approximate distribution of the sample mean, $\bar{X}_{72}$.

   ii) Find the probability that $\bar{X}_{72}$ is greater than 6.5.

5. $X \sim \text{B}(10, 0.3)$. A random sample of 80 observations of X is taken.

   i) State the approximate distribution of the sample mean, $\bar{X}_{80}$.

   ii) Find the probability that $\bar{X}_{72}$ is greater than 3.25.

6.8 Descriptions of some sampling methods

Note: The details of sampling methods are not needed for this course, but thinking about why different methods have been developed over time may help you to understand more about the difficulties in obtaining a good sample to work with.

Systematic (or purposive, or periodic) sampling: Choose a number at random from 1 to n, and then take every nth value thereafter. This means that all units have an equal chance of being selected, but it does **not** constitute a random sample, as not all combinations are equally likely. There is a particular danger with this method, especially for 'production line sampling' as any faults with a particular machine in the system will affect items regularly, and the systematic method of choosing the items to be sampled means that you are very likely either to miss them all or hit them all (or half or a third of them).

Stratified sampling: This is where the population has identifiable groups (strata) within it for which we have reason to believe that the behaviour will be different. The sample then is made up of the same proportions of each group. [It is very effective in reducing the variability of sampling estimates when the population has known strata because it removes one source of variation, since the numbers out of each group are the same every time the sample is carried out. However, you can only use it when the size of the groups in the population are known].

Quota sampling: This is where the sample used is required to have specified proportions (the quotas) of different groups, for example of men/women, adults/children, different social classes etc. The sampler has more freedom than in a stratified (which it resembles in certain ways), a systematic or a random sample since in all of these the system determines the individuals to be taken, and consequently there is the danger that subjective judgements will introduce bias in the sample. It is much, much easier to accomplish than the 'proper statistical method' it mimics i.e. stratified, so it is possible to use rather larger samples (a good feature) and if experienced interviewers are used, and care taken in structuring the method so as to avoid predictable causes of bias, then it is practically useful (cheap and cheerful is fine when you only need a rough idea of what is happening, particularly if you would like to know it quickly).

N.B. Stratified and quota sampling are both **'representative'** samples in that the profile of the sample should broadly match the profile of the population in respect of the groupings chosen.

Cluster sampling: If (and it is a big if) a population naturally sub-divides into a number of small *homogeneous* groups or *clusters* then it can be convenient to select a number of clusters and sample from these clusters (or even examine every member). Clusters might well be generated on a geographical basis - and **would significantly reduce the costs of collecting the sample information in this case**. Political opinion polls use a variation on this, where a number of areas over the country are chosen, and a sample from each area is taken, with a total sample size of around 1000 people being used commonly.

Self selection: However bad any of these methods are - even poor quota samples - they are much better than self-selecting samples e.g. radio or television 'phone-ins' where listeners record their 'vote'. Views of those who feel strongly about something are much more frequently expressed, and then the negative strong feelings are usually over-represented even within this group, than the apathetic majority.

Comparisons of methods:

The choice of sampling method and size of sample to be used is essentially a pragmatic matter balancing cost, speed of availability of the analysis, and the quality of the information which can be expected.

If stratification is well defined then stratified sampling is the 'best approach' statistically, and the size of the sample is chosen so that the different strata can be represented proportionally. Within strata, the sample should ideally be taken at random, so that the question of potential bias is not an issue. With other types of sampling you have to ask whether there is any likelihood that the sample obtained may not be representative of the population as a whole. The reasons for using other types is practical - the extra costs involved in stratified sampling will often mean that you can afford to take a larger sample size with another method and end up with better quality of information at less cost overall.

Note that all of the above discussion is on the assumption that you are taking a small sample in comparison with the size of the population, so that you can neglect the effects of sampling proportion. Until you are sampling a significant proportion (e.g. 50%) of the population the effect of the sample proportion is much smaller as an effect than the sample size.

Methods of collection: Data is increasingly collected automatically by electronic measurement devices that are programmed to take measurements regularly, and record them e.g. in a hospital situation, or a production process. However, not all types of data can be collected in this way, and there is still a large amount of data collected through surveys, using questionnaires, and by direct observation and recording. In designing questionnaires, care needs to be taken that questions are unambiguous, and do not use language which will bias the respondent's answer. They should be as simple to understand and answer as possible to collect the required information, and thought is needed about how the information is to be analysed to ensure that the data is collected in a form which makes the analysis easily accessible.

Summary exercise 6

1. Karolina wants to choose one student at random from Annika, Belinda and Charlize. She looks in the sports results of Monday's paper. If the first match is a home win she will choose, Annika. If it is a draw she will choose Belinada and if it is an away win then she will choose Chalize.

   Explain why this is not a fair method of choosing the student.

2. **a)** Explain what you understand by the term *random sample*

   b) Give an example of bias in sampling.

3. A manufacturer of rechargeable batteries wants to know how long they will operate for before needing to be charged.

   Explain why a sample should be used rather than a census.

4. Using the random number generator on your calculator, take a simple random sample of size 10 from the table below, and calculate the mean of your sample.

   79 45 79 56 76 62 64 56 44 48 75 40 38 67 37
   50 77 55 35 39 61 73 39 70 36 70 41 54 35 72
   48 62 42 39 35 36 77 62 55 54 48 44 54 74 53
   63 80 80 72 63 80 80 58 63 41 61 48 39 70 63
   63 65 41 44 75 40 78 67 58 70 55 71 52 42 69

   There are 75 values, with 15 per row.

5. A bag contains two 50 cent coins, a 20 cent coin and a 10 cent coin. Two coins are to be randomly chosen.

   a) If the coins are marked as A, B, C and D, list the twelve possible outcomes of taking one coin and then a second coin.

   b) Hence find the sampling distribution of the mean value of the coins taken

6. X is a random variable with mean 22.4 and variance 7.9.

   a) Find the variance of the mean of a sample of 20 independent observations of X.

   b) What is the minimum size of sample needed in order that the variance of the sample mean is less than 1?

7. The contents of bottles of fruit juice are normally distributed with mean 505 millilitres and standard deviation 6.1 millilitres.

   a) Give the distribution of the mean content of a random sample of 6 bottles.

   b) Find the probability that the mean content of a random sample of 6 bottles is less than 500 millilitres.

8. Packets of sweets are labelled as containing 250 grams. The actual contents can be modelled by a Normal distribution with mean 252 grams and standard deviation 3.5 grams.

   a) Find the probability that the mean contents of a random sample of 10 packets is at least 250 grams.

   b) What is the smallest size of random sample for which there is probability of under 0.5% that the mean contents is under 250 grams?

9 Marie wants to choose one student at random from Anthea, Bill and Charlie. She throws two fair coins. If both coins show tails she will choose Anthea. If both coins show heads she will choose Bill. If the coins show one of each she will choose Charlie.

   a) Explain why this is not a fair method for choosing the student. [2]

   b) Describe how Marie could use the two coins to give a fair method for choosing the student. [2]

Cambridge International A Level and AS Mathematics 9709 Paper 71 Q1 June 2013

10. a) Write down the mean and variance of the distribution of the means of random samples of size n taken from a very large population having mean μ and variance σ^2. [2]

   b) What, if anything, can you say about the distribution of sample means

     i) if n is large, [1]

     ii) if n is small? [1]

Cambridge International A Level and AS Mathematics 9709 Paper 7 Q2 November 2006

Chapter summary

- You can use a sample to gather information about a sample when a census would be impractical.
- Samples need to be constructed carefully to avoid bias.
- A sampling frame is a list of all of the sampling units or elements (often people) in the population to be samples.
- A statistic is a function that depends only on the observed values in a sample i.e. it must not use any of the parameter values.
- The probability distribution is known as the sampling distribution of the statistic.
- Random numbers can be used to draw a random sample from a population.
- The Central Limit Theorem says that the distribution of sample means becomes approximately Normal as the sample size increases, no matter what the underlying population is: $n = 30$ is applied as a standard rule of thumb as to what sample size is needed in order to use the Central Limit Theorem.

Review Exercise B

1. Alan wishes to choose one child at random from the eleven children in his music class. The children are numbered 2, 3, 4, and so on, up to 12. Alan then throws two fair dice, each numbered from 1 to 6, and chooses the child whose number is the sum of the scores on the two dice.

    a) Explain why this is an unsatisfactory method of choosing a child. [2]

    b) Describe briefly a satisfactory method of choosing a child. [2]

    Cambridge International AS and
    A Level Mathematics 9709 Paper 7
    Q1 November 2008

2. A continuous random variable X has probability density function given by

    $$f(x) = \begin{cases} a + \frac{1}{3}x & 1 \le x \le 2, \\ 0 & \text{otherwise,} \end{cases}$$

    where a is a constant.

    a) Show that the value of a is $\frac{1}{2}$. [3]

    b) Find $P(X > 1.8)$. [2]

    c) Find $E(X)$. [3]

    Cambridge International AS and
    A Level Mathematics 9709 Paper 7
    Q5 November 2005

3.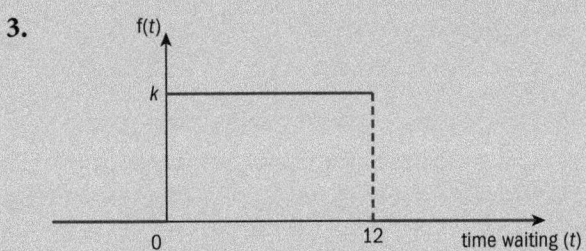

Fred arrives at random times on a station platform. The times in minutes he has to wait for the next train are modelled by the continuous random variable for which the probability density function f is shown above.

a) State the value of k. [1]

b) Explain briefly what this graph tells you about the arrival times of trains. [1]

Cambridge International AS and
A Level Mathematics 9709 Paper 71
Q1 June 2010

4. Over a long period of time it is found that the amount of sunshine on any day in a particular town in Spain has mean 6.7 hours and standard deviation 3.1 hours.

    a) Find the probability that the mean amount of sunshine over a random sample of 300 days is between 6.5 and 6.8 hours. [4]

    b) Give a reason why it is not necessary to assume that the daily amount of sunshine is normally distributed in order to carry out the calculation in part a). [1]

    Cambridge International AS and
    A Level Mathematics 9709 Paper 7
    Q2 November 2004

5. Climbing ropes produced by a manufacturer have breaking strengths which are normally distributed with mean 160 kg and standard deviation 11.3 kg. A group of climbers have weights which are nomally distributed with mean 66.3 kg and standard deviation 7.1 kg.

a) Find the probability that a rope chosen randomly will break under the combined weight of 2 climbers chosen randomly. [5]

Each climber carries, in a rucksack, equipment amounting to half his won weight.

b) Find the mean and variance of the combined weight of a climber and his rucksack. [3]

c) Find the probability that the combined weight of a climber and his rucksack is greater than 87 kg. [2]

Cambridge International AS and A Level Mathematics 9709 Paper 7 Q5 November 2006

6. A random variable X has the distribution Po(1.6).

a) The random variable R is the sum of three independent values of X. Find $P(R < 4)$. [3]

b) The random variable S is the sum of n independent values of X. It is given that

$$P(S = 4) = \frac{16}{3} \times P(S = 2).$$

Find n. [4]

c) The random variable T is the sum of 40 independent values of X. Find $P(T > 75)$. [4]

Cambridge International AS and A Level Mathematics 9709 Paper 71 Q7 November 2012

7. At a town centre car park the length of stay in hours in denoted by the random variable X, which has probability density function given by

$$f(x) = \begin{cases} kx^{-\frac{3}{2}} & 1 \le x \le 9, \\ 0 & \text{other wise,} \end{cases}$$

where k is a constant.

a) Interpret the inequalities $1 \le x \le 9$ in the definition of $f(x)$ in the context of the question. [1]

b) Show that $k = \frac{3}{4}$. [2]

c) Calculate the mean length of stay. [3]

The charge for a length of stay of x hours is $(1 - e^{-x})$ dollars.

d) Find the length of stay for the charge to be at least 0.75 dollars [3]

e) Find the probability of the charge being at least 0.75 dollars. [2]

Cambridge International AS and A Level Mathematics 9709 Paper 7 Q7 November 2006

8. When Sunil travels from his home in England to visit his relatives in India, his journey is in four stages. The times, in hours, for the stages have independent normal distributions as follows.

Bus from home to the airport: N(3.75, 1.45)

Waiting in the airport: N(3.1, 0.785)

Flight from England to India: N(11, 1.3)

Car in India to relatives: N(3.2, 0.81)

a) Find the probability that the flight time is shorter than the total time for the other three stages. [6]

b) Find the probability that, for 6 journeys to India, the mean time waiting in the airport is less than 4 hours. **[3]**

Cambridge International AS and A Level Mathematics 9709 Paper 71 Q6 June 2009

9. The queuing time, T minutes, for a person queuing at a supermarket checkout has probability density function given by

$$f(t) = \begin{cases} ct(25 - t^2) & 0 \le t \le 5, \\ 0 & \text{otherwise,} \end{cases}$$

where c is a constant.

a) Show that the value of c is $\dfrac{4}{625}$. **[3]**

b) Find the probability that a person will have to queue for between 2 and 4 minutes. **[3]**

c) Find the mean queuing time. **[4]**

Cambridge International AS and A Level Mathematics 9709 Paper 7 Q7 June 2004

10. A random variable X has probability density function given by

$$f(x) = \begin{cases} 1 - \dfrac{1}{2}x & 0 \le x \le 2, \\ 0 & \text{otherwise.} \end{cases}$$

a) Find $P(X > 1.5)$. **[2]**

b) Find the man of X. **[2]**

c) Find the median of X. **[3]**

Cambridge International AS and A Level Mathematics 9709 Paper 7 Q4 June 2003

11. a)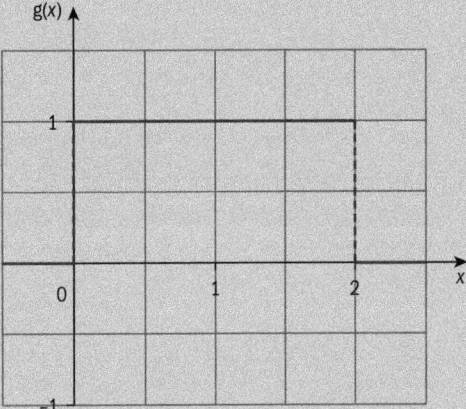

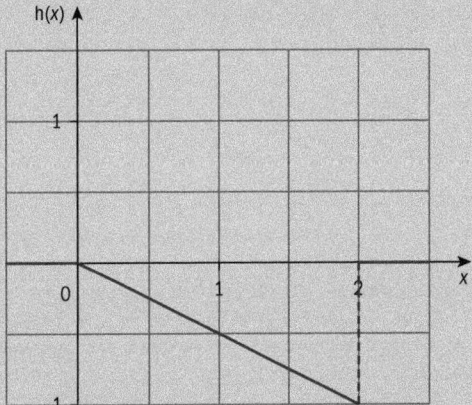

The diagrams show the graphs of two functions, g and h. For each of the functions g and h, give a reason why it cannot be a probability density function. **[2]**

b) The distance, in kilometres, travelled in a given time by a cyclist is represented by the continuous random variable X with probabiltiy density function given by

$$f(x) = \begin{cases} \dfrac{30}{x^2} & 10 \le x \le 15, \\ 0 & \text{otherwise.} \end{cases}$$

i) Show that $E(X) = 30 \ln 1.5$. **[3]**

ii) Find the median of X. Find also the probability that X lies between the median and the mean. [5]

Cambridge International AS and A Level Mathematics 9709 Paper 71 Q4 June 2011

12. The random variable X has the distribution $B(10, 0.15)$. Find the probability that the mean of a random sample of 50 observations of X is greater than 1.4. [5]

Cambridge International AS and A Level Mathematics 9709 Paper 7 Q1 June 2007

13. Random samples of size 120 are taken from the distribution $B(15, 0.4)$.

   a) Describe fully the distribution of the sample mean. [3]

   b) Find the probability that the mean of a random sample of size 120 is greater than 6.1. [3]

Cambridge International AS and A Level Mathematics 9709 Paper 7 Q3 June 2006

14. The lengths of time people take to complete a certain type of puzzle are normally distributed with mean 48.8 minutes and standard deviation 15.6 minutes. The random variable X represents the time taken in minutes by a randomly chosen person to solve this type of puzzle. The times taken by random samples of 5 people are noted. The mean time $\bar{X}$ is calculated for each sample.

   a) State the distribution of $\bar{X}$, giving the values of any parameters. [2]

   b) Find $P(\bar{X} < 50)$. [3]

Cambridge International AS and A Level Mathematics 9709 Paper 7 Q2 June 2008

15. Jenny has to do a statistics project at school on how much pocket money, in dollars, is received by students in her year group. She plans to take a sample of 7 students from her year group, which contains 122 students.

   a) Give a suitable method of taking this sample. [1]

   Her sample gives the following results.

   13.40 10.60 26.50 20.00 14.50 15.00 16.50

   b) Find unbiased estimates of the population mean and variance. [3]

   c) Is the estimated population variance more than, less than or the same as the sample variance? [1]

   d) Describe what you understand by 'population' in this question. [1]

Cambridge International AS and A Level Mathematics 9709 Paper 7 Q2 June 2005

16. A magazine conducted a survey about the sleeping time of adults. A random sample of 12 adults was chosen from the adults travelling to work on a train.

   a) Give a reason why this is an unsatisfactory sample for the purposes of the survey. [1]

b) State a population for which this sample would be satisfactory. [1]

A satisfactory sample of 12 adults gave numbers of hours of sleep as shown below.

4.6 6.8 5.2 6.2 5.7 7.1
6.3 5.6 7.0 5.8 6.5 7.2

c) Calculate unbiased estimates of the mean and variance of the sleeping times of adults. [3]

Cambridge International AS and A Level Mathematics 9709 Paper 7 Q1 June 2008

17. The random variable X has probability density function given by

$$f(x) = \begin{cases} \dfrac{k}{(x+1)^2} & 0 \le x \le 1, \\ 0 & \text{otherwise,} \end{cases}$$

where k is a constant.

a) Show that $k = 2$. [2]

b) Find a such that $P(X < a) = \dfrac{1}{5}$. [3]

c)

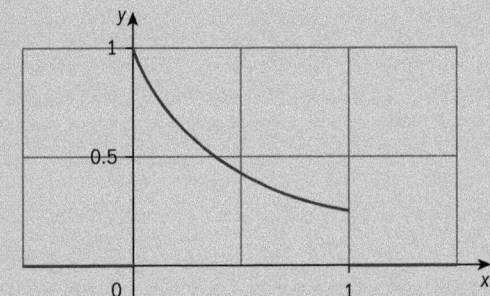

The diagram shows the graph of $y = f(x)$. The median of X is denoted by m. Use the diagram to explain whether $m < 0.5$, $m = 0.5$ or $m > 0.5$. [2]

Cambridge International AS and A Level Mathematics 9709 Paper 71 Q4 June 2012

Maths in real-life

Modelling statistics

The Golden Gate Bridge is one of the world's iconic bridges as it spans 1,300 metres and was the longest suspension bridge in the world from when it opened in 1937 until 1964. When it opened, during the first fiscal year it averaged just over 9000 vehicles a day, which rose steadily over the next 50 years to peak at nearly 120,000 vehicles a day, since then it has reduced slightly but still averages over 100,000 vehicles a day. The bridge has a moveable median barrier which is moved several times during the day to optimize traffic flows – on weekday mornings more traffic is heading south into San Francisco and 4 lanes are open south and two north. In the afternoons it is 4 lanes north and two south while at other times on weekdays and at weekends there are 3 lanes open in each direction. The bridge operators have now introduced congestion price charging in order to try to reduce the traffic loads at peak traffic times.

The statistical challenges for building and maintaining bridges are considerable – the structural specifications given to the architects and engineers in building bridges are based on projected uses in the future – the Golden Gate Bridge celebrated its 75th anniversary in 2012 and for about 40 years it has carried more than ten times the number of vehicles it carried when it opened. It is remarkable that the structure was designed to be strong enough to cope with such a large increase in usage over its lifetime.

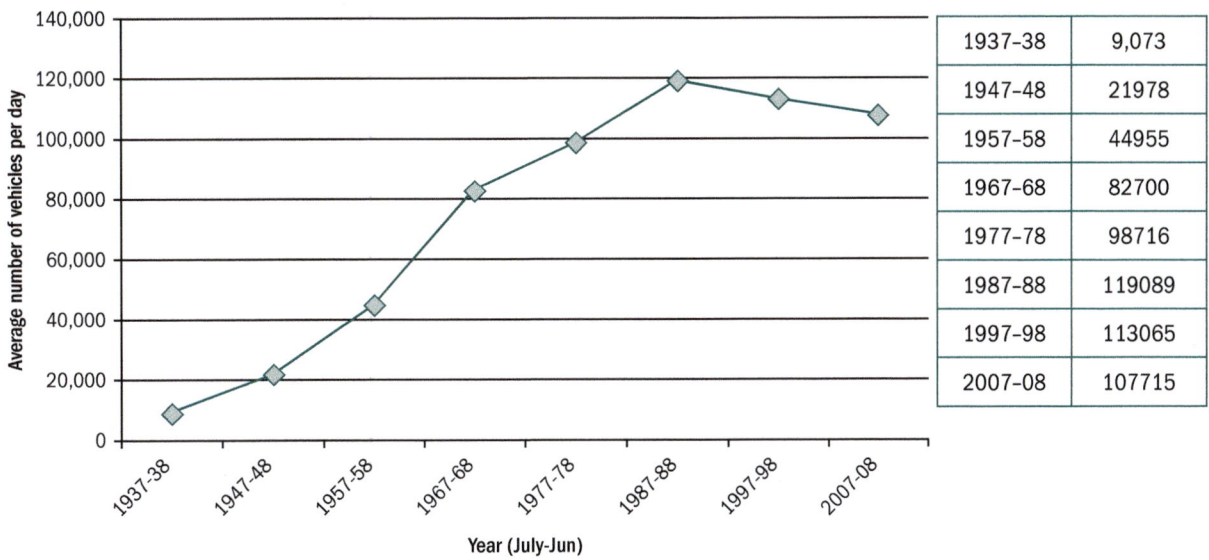

Traffic on the Golden Gate Bridge

Year (July–Jun)	Average number of vehicles per day
1937–38	9,073
1947–48	21978
1957–58	44955
1967–68	82700
1977–78	98716
1987–88	119089
1997–98	113065
2007–08	107715

People are creatures of habit, so if traffic layouts are to be changed on the bridge at different times of the day, there is considerable virtue in keeping the structure as simple as possible – but the operators have found that while the morning commute is fairly predictable, the afternoon northbound commute is much more variable. There has also been an increase in southbound traffic in the afternoons, meaning that fixed times for the lane changes was resulting in considerable delays for southbound traffic. The current model is to allocate traffic lanes to accommodate northbound traffic without introducing unacceptable delays southbound. Tolls collected on the Golden Gate Bridge now exceed $100 million annually and about half of the tolls collected are used to subsidise other methods of transport (the Golden Gate Ferry and Golden Gate Transit) which are estimated to reduce the commute traffic by around 25%. Decisions about pricing tolls, and fares for other forms of transport are complex, involving sophisticated economic and statistical modelling, as the Golden Gate Bridge tries to manage to maintain its iconic status in its fourth quarter century operations.

A haulage firm has to quote prices for jobs in which one of their major costs is the length of time it will take – they have to quote in advance, sometimes quite a long time in advance, but don't know how long the journey will take until it happens. Having some idea of what constitutes a typical journey – using a range of values rather than just a single point estimate – helps the company make sensible economic pricing decisions.

Objectives

After studying this chapter you should be able to:

- Calculate unbiased estimates of the population mean and variance from a sample, using either raw or summarised data (only a simple understanding of the term 'unbiased' is required);

- Determine a confidence interval for a population mean in cases where the populations is normally distributed with known variance or where a large sample is used;

- Determine, from a large sample, an approximate confidence interval for a population proportion.

Before you start

You should know how to:

1. Calculate the mean and variance of a set of data

    Calculate the mean and variance of

    3 5 6 6 8 9 12 13 15 15 16

    $n = 11, \sum x = 108; \sum x^2 = 1270$

    $\bar{x} = \dfrac{108}{11} = 982; \mathrm{Var}(x) = \dfrac{1270}{11} - \bar{x}^2 = 19.1$

Skills check:

1. Calculate the mean and variance of

    5 5 6 8 9 10 11 11 12 14 17 26

7.1 Interval estimation

If you want to estimate the mean (μ) of a population by taking a sample, the best you can do is to use the sample mean $\bar{x}$. We will define more closely what we mean by 'best' in the next section.

If you look at the following three samples (each of 15 values), the sample means are each 44.82 and you would use that value as the point estimate of each of the population means. However, would you feel that you have the same quality of information in the three cases?

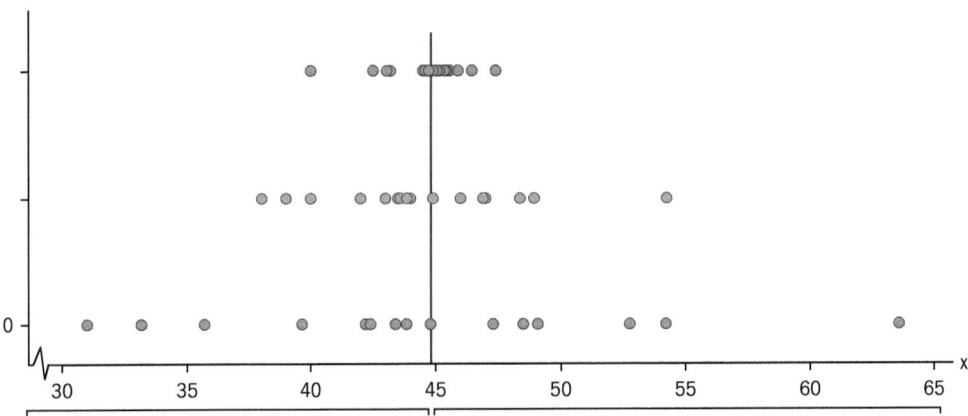

Top	middle	bottom
43.05	39.08	30.99
44.73	40.39	33.17
45.58	43.88	35.77
44.89	46.15	39.33
45.13	44.43	42.37
45.69	38.00	42.77
47.48	47.08	43.44
44.67	42.17	43.86
43.72	48.91	44.80
44.59	43.97	47.30
46.44	44.90	48.48
42.64	46.99	49.15
43.11	43.69	52.81
45.55	54.31	54.43
44.99	48.37	63.62
44.82	44.82	44.82

The values in the top sample are very close together where the ones in the bottom sample have much more variability. Instead of just giving a point estimate which loses all the extra information we can see from the spread of the sample values, we are going to develop interval estimates which capture something about how reliable the estimate is.

We will get to the formal construction of the estimates in later sections of this chapter, but informally it seems reasonable that the interval estimate for the top sample would be the smallest width, and the estimate based on the bottom sample would have the largest width.

One more thing to consider when we think of how reliable our information is: suppose we only had a few values rather than the 15 in each of the samples above.

The set of 15 is the middle set of data shown above and the small data set shows the same amount of variability as it has. Again, informally, the extra information from the larger sample should make you feel the estimate is more reliable for the larger sample – and the interval estimate for the larger sample will be narrower to reflect the better quality of information.

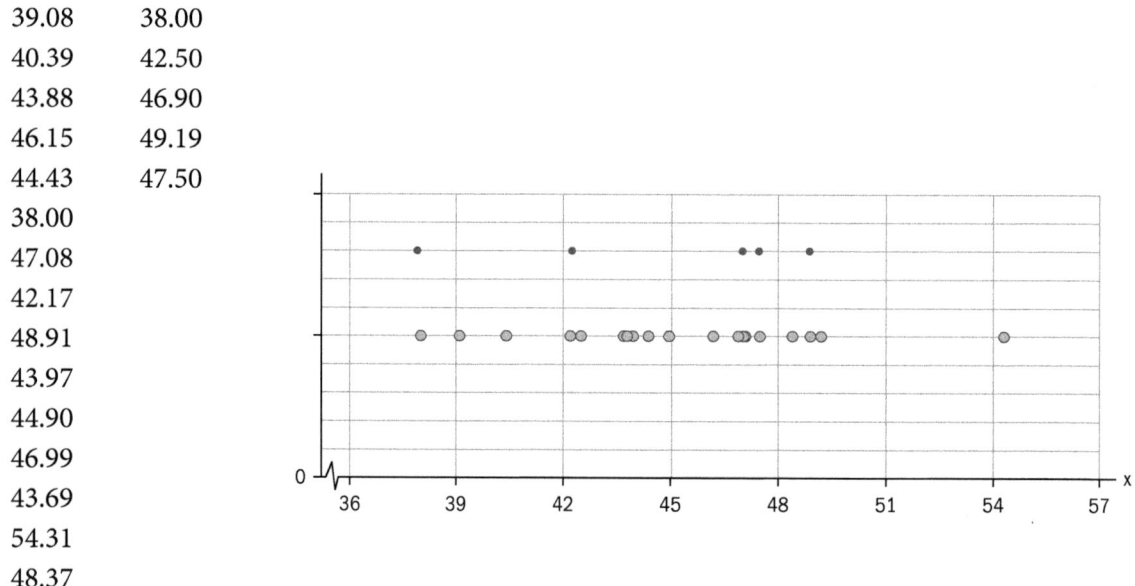

39.08	38.00
40.39	42.50
43.88	46.90
46.15	49.19
44.43	47.50
38.00	
47.08	
42.17	
48.91	
43.97	
44.90	
46.99	
43.69	
54.31	
48.37	

Before you learn how to construct **confidence intervals** formally (this is what the interval estimates are called), you need to meet some other important ideas in estimation.

There is no exercise for this section.

7.2 Unbiased estimate of the population mean

In chapter 3, you saw:

$$E(\bar{X}) = E\left(\frac{1}{n}\sum_{i=1}^{n} X_i\right) = \frac{n\mu}{n} = \mu$$

This says that 'on average' the sample mean will give the true value of the population mean. This is the definition of an **unbiased estimator** – of any parameter. But there are many 'unbiased estimates' of the mean: for example X_1, X_n, $\frac{X_1 + X_2}{2}$ are all also unbiased estimates of the population mean. So what else would we want in an estimator? If it gave estimates close to the true value more often, it would be doing a better job as an estimator than one where the estimates were more widely spread. Consider the variances of the estimators listed here:

$$\text{Var}(\bar{X}) = \text{Var}\left(\frac{1}{n}\sum_{i=1}^{n} X_i\right) = \frac{n\sigma^2}{n^2} = \frac{\sigma^2}{n}$$

$$\text{Var}(X_1) = \sigma^2$$

$$\text{Var}(X_2) = \sigma^2$$

$$\text{Var}\left(\frac{X_1 + X_2}{2}\right) = \frac{\sigma^2}{2}$$

So $\bar{X}$ is the unbiased estimator out of this group with the lowest variance. In fact, it is the estimator with the lowest variance out of all unbiased estimators – but the proof of that is beyond this course. What does it mean in practice? The table below shows the results of a simulation taking observations from a Normal distribution with mean 50 and standard deviation 5.

Statisticians talk about the **efficiency** of an estimator – or usually the relative efficiency as a way of comparing two estimators – using the variability of the estimator as the basis for this measure, but this goes beyond the scope of this course.

X_1	X_2	X_3	X_4	X_5	X_6	X_7	X_8	$\bar{X}$	X_1	X_2	$\frac{X_1 + X_2}{2}$
53.2	54.0	54.9	60.2	51.3	50.8	46.4	47.4	52.3	53.2	54.0	53.6
52.6	59.7	44.6	46.9	53.2	48.4	45.9	47.7	49.9	52.6	59.7	56.1
51.1	49.7	44.6	53.5	51.5	55.9	57.2	50.3	51.7	51.1	49.7	50.4
56.5	55.0	49.5	56.1	55.0	43.5	50.8	51.7	52.3	56.5	55.0	55.8
51.3	51.5	55.4	48.1	50.5	41.3	49.6	55.8	50.5	51.3	51.5	51.4
41.4	47.8	49.1	50.4	55.7	45.4	51.9	50.7	49.1	41.4	47.8	44.6
55.0	49.8	39.2	53.9	47.4	52.3	54.5	56.2	51.1	55.0	49.8	52.4
47.2	45.4	58.4	52.4	51.3	53.7	49.4	46.7	50.5	47.2	45.4	46.3
46.5	48.6	54.6	46.8	46.4	44.4	55.6	46.2	48.6	46.5	48.6	47.6
44.4	42.8	52.0	46.6	58.4	47.0	53.7	61.8	50.8	44.4	42.8	43.6
40.4	53.9	45.3	52.9	45.6	50.6	50.8	49.4	48.6	40.4	53.9	47.2
57.3	51.2	57.2	53.4	53.4	47.0	48.0	49.7	52.2	57.3	51.2	54.3
56.5	47.6	51.8	45.9	46.0	53.5	51.7	46.1	49.9	56.5	47.6	52.1
50.2	41.2	51.0	46.4	54.1	51.0	44.6	51.8	48.8	50.2	41.2	45.7
42.8	44.8	58.5	45.9	53.3	59.0	49.2	41.8	49.4	42.8	44.8	43.8

From the graphic you can see that $\overline{X}$ returns estimates for the mean which consistently stay close to the true value of 50. $\frac{X_1 + X_2}{2}$ is 'better' than just the single values but $\overline{X}$ is much better.

Although it is not part of the requirements for this course it is worth mentioning another property of $\overline{X}$ as an estimator for the population mean. It is **consistent** – which means that as the sample size increases, the probability that the estimator is close to the true value also increases. Consistency is a property that intuitively feels desirable in an estimator: if you were to take a huge sample (of several million observations say), you would want your estimate to be really close to the true value – that's all that consistency means.

Exercise 7.2

1. The table below shows a sample of 30 observations taken from a Normal distribution with mean μ and standard deviation 10.

    Calculate the following unbiased estimators of μ:

    a) the mean of all 30 values

    b) the mean of each block of 5 values (will give 6 separate estimates)

    c) the mean of each row of 10 values (will give 3 separate estimates)

82.1	51.9	59.9	66.2	71.5		69.0	59.1	50.5	73.2	64.4
54.2	81.6	57.3	53.1	57.5		63.9	62.3	78.5	74.7	58.7
57.6	68.9	61.0	67.4	45.8		53.0	70.2	48.6	81.5	46.5

2. Repeat Q1 with the set of data shown below

75.6	59.4	48.5	65.3	73.7		61.8	65.7	64.1	64.3	60.1
50.0	78.7	51.5	61.2	54.9		62.3	62.3	56.1	60.9	56.9
67.5	57.4	60.3	52.9	59.9		76.7	56.2	71.0	48.8	49.5

3. On the website you will find a spreadsheet file named Exercise 7_2 Q3. It was used to generate the data in the two questions above.

    The mean, μ, of the population being sampled is actually 60 and it is important to realize that the best estimator will not always give the value closest to 60 – especially when you have multiple bites at the cherry, for example in taking 6 estimates each based on 5 observations.

    The spreadsheet is set up to show you the comparable estimates from parts a) – c) of questions 1 and 2 – without you doing the tedious work of calculating all the averages. Using the spreadsheet to generate a number of samples, explore how variable the estimators in parts a (based on 30 values), b (based on 5 values) and c (based on 10 values) are.

Pressing F9 will produce a new set of 30 values and also all the estimates of the mean. The tedious work has been taken away – but you need to think about what you see – not just keep pressing F9 to see a new sample. It may help you build up the feeling if you make a note of the overall sample values you see because there is only one per simulation and it is hard to keep the values in your head.

7.3 Unbiased estimate of the population variance

We know that the variance of a random variable can be calculated as $E\{(X -\mu)^2\}$ and that this is equivalent to $E\{(X)^2 -\mu^2\}$.

If somehow we knew what μ was before we took a sample, we would be able to use it in a calculation to estimate the variance of the population. However, that situation is extremely rare and you normally have to use $\overline{X}$ instead of μ in any estimation process. Algebraically, it is possible to prove that the value of $\sum(x_i - k)^2$ is a minimum when $k = \overline{x} = \frac{1}{n}\sum x_i$, but we also know that $\overline{x}$ is not always equal to the true value μ.

This means that $\frac{1}{n}\sum(x_i - \overline{x})^2$ which is the variance of the sample will tend to underestimate the variance because it is using the centre of the data $(\overline{x})$ rather than the population mean μ.

However, it can be shown that

$$s^2 = \frac{1}{n-1}\sum(x_i - \overline{x})^2 = \frac{1}{n-1}\left(\sum x^2 - n\overline{x}^2\right) = \frac{1}{n-1}\left(\sum x^2 - \frac{(\sum x)^2}{n}\right),$$

is an unbiased estimator of the population variance. The first form of this expression is the way it is defined, and the second and third forms are computational equivalents which are often easier to use.

The proof is not part of this course, but it is not too difficult, and it is included at the end of this section.

Advice on calculator use
Most calculators will give two standard deviations – normally labeled as σ_n and σ_{n-1}, or as σ and s – the first is used if you have a full set of data – i.e. it is a population, and the second should be used if you have a sample from a larger population.

Note: while s^2 is an unbiased estimator of the variance, s is **not** an unbiased estimator of the standard deviation, and there is no estimator which would provide an unbiased estimate for all distributions because the nature of the bias depends on the particular distribution. s is an underestimate of the standard deviation, but not in the proportional way that s^2 behaved where we can make a simple adjustment to obtain an unbiased estimator.

For a set of data in grouped intervals you saw in S1 that the variance was given by $\frac{\sum(x_i - \overline{x})^2 f_i}{\sum f_i}$ or $\frac{\sum x_i^2 f_i}{\sum f_i} - \overline{x}^2$. If the data is a sample rather than the full population then you need to adjust as above and use

$$s^2 = \frac{\sum(x_i - \overline{x})^2 f_i}{(\sum f_i)-1} = \left(\frac{n}{n-1}\right)\left\{\frac{\sum x_i^2 f_i}{\sum f_i} - \overline{x}^2\right\} \text{ where } n = \sum f_i.$$

Because you take all values in an interval at the mid-interval value, this is no longer an unbiased estimate of the population variance (because variance takes squared distances from the mean) but it is the best estimate available so it is the one that is used.

Example 1

The number of passengers travelling on an executive jet on a random sample of 10 journeys is given below.
Find unbiased estimates of the mean and variance of the number of passengers on a journey in the executive jet.

5, 2, 8, 3, 3, 5, 1, 3, 0, 4

$\sum x = 34$; $\sum x^2 = 162$, so $\bar{x} = \dfrac{34}{10} = 3.4$; $s^2 = \dfrac{1}{9}\left(162 - \dfrac{34^2}{10}\right) = \dfrac{46.4}{9} = 5.16$
are the unbiased estimates of the mean and variance.

Example 2

The weight in kg (to the nearest kg) of the hand luggage of a random sample of 10 passengers travelling on a long haul flight is given below.

Find estimates of the mean and variance of the weight of hand luggage for passengers on the flight.

8, 7, 9, 10, 10, 8, 4, 7, 9, 10

Here the weights are actually intervals: 9 kg is between 8.5 and 9.5 kg so although this looks like the data in example 1, the estimate of the variance here is not an unbiased estimator.

$\sum x = 82$; $\sum x^2 = 704$, so $\bar{x} = \dfrac{82}{10} = 8.2$ $s^2 = \dfrac{1}{9}\left(704 - \dfrac{82^2}{10}\right) = \dfrac{31.6}{9} = 3.51$

are the best available estimates of the mean and variance.

Example 3

The weight, x kg, of a random sample of suitcases checked in on a long haul flight is given below. Find estimates of the mean and standard deviation of the weight of suitcases checked in for the flight.

Weight (kg)	$4.5 \leq x < 9.5$	$9.5 \leq x < 14.5$	$14.5 \leq x < 19.5$	$19.5 \leq x < 24.5$
frequency	5	9	35	24

The interval midpoints are 7, 12, 17 and 22,

$\sum f = 73$; $\sum mf = 1266$; $\sum m^2 f = 23272$, so

$\bar{x} = \dfrac{1266}{73} = 17.3$; $s^2 = \dfrac{1}{72}\left(23272 - \dfrac{1266^2}{73}\right) = \dfrac{1316.4}{72} = 18.28387...$

$\Rightarrow s = \sqrt{18.28387..} = 4.28$

Proof that $s^2 = \dfrac{1}{n-1}\sum(x_i - \bar{x})^2$ **is an unbiased estimator of the population variance**

Note: Deriving this proof is not required by the syllabus.

The proof makes use of a number of relationships which you have met previously:

$$\sigma^2 = \mathrm{Var}(X) = \mathrm{E}\{(X - \mu)^2\} = \mathrm{E}(X^2) - \mu^2 \Rightarrow \mathrm{E}(X^2) = \mu^2 + \sigma^2 \quad \text{see page 100 of S1}$$

$$\mathrm{Var}(\bar{X}) = \frac{\sigma^2}{n} = \mathrm{E}\{(\bar{X} - \mu)^2\} = \mathrm{E}(\bar{X}^2) - \mu^2 \Rightarrow \mathrm{E}(\bar{X}^2) = \mu^2 + \frac{\sigma^2}{n} \quad \text{see chapter 3 of S2}$$

$$\sum(x_i - \bar{x})^2 = \sum x_i^2 - n\bar{x}^2 \quad \text{see page 22 of S1 (multiplied through by } n)$$

Then the result for the unbiased estimator follows quite quickly:

$$\mathrm{E}\left\{\sum(x_i - \bar{x})^2\right\} = \mathrm{E}\left\{\sum x_i^2 - n\bar{x}^2\right\} = \mathrm{E}\left\{\sum x_i^2\right\} - n\mathrm{E}\{\bar{x}^2\}$$

$$= n(\mu^2 + \sigma^2) - n\left(\mu^2 + \frac{\sigma^2}{n}\right)$$

$$= n\mu^2 + n\sigma^2 - n\mu^2 - \sigma^2 = (n-1)\sigma^2$$

So $\mathrm{E}\left\{\dfrac{1}{n-1}\sum(x_i - \bar{x})^2\right\} = \sigma^2$

While using the mid-interval value for grouped frequency data is the best you can do, it is worth being aware of quite common situations where the best is really not all that good. When data tails off considerably at one or both ends it is common for unequal intervals to be used – wider intervals where not much is going on makes it less likely that you try to over-interpret sparse data, and narrow intervals where the data is dense so you can see more detail there. The wide intervals cause considerable inaccuracy to be introduced in estimating the variance and standard deviation, but where the distribution is close to symmetrical, the estimate of the mean is likely to be reasonably accurate. However, if the distribution is heavily skewed, the estimate of the mean is going to be affected as well as the estimates of measures of spread.

Consider the following histogram, of a data set which is a large sample of household incomes, in thousands of dollars. The data was grouped in intervals of widths 20, 20, 20, 40 and 100 thousand dollars and it is heavily skewed.

The estimate of the mean income is $40,700, with an estimated standard deviation of $33,200.

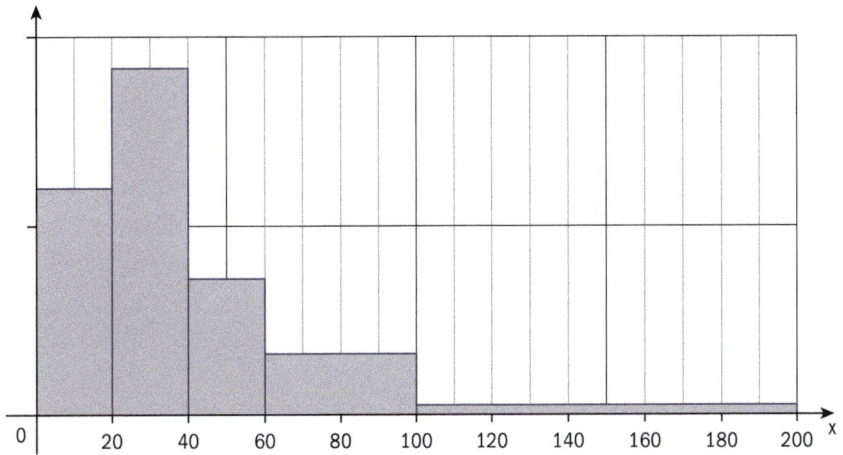

If you model the data to estimate the frequencies in intervals of constant width to replace the top two wide intervals, it looks something like this:

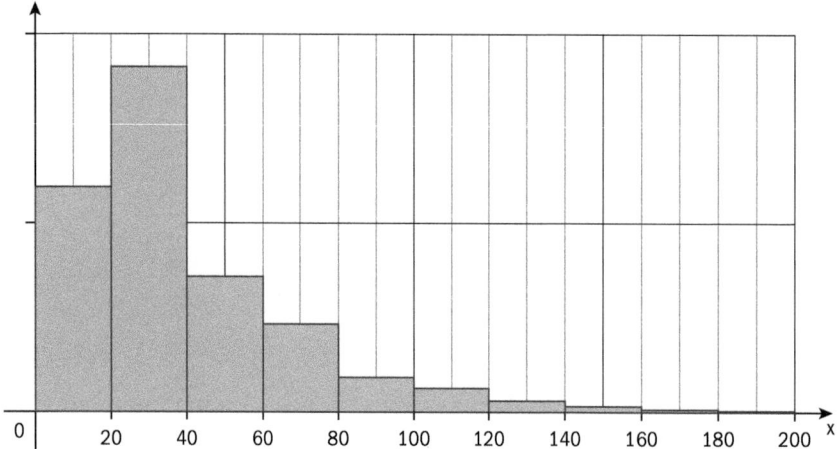

The estimate of the mean income is now $38,900, with an estimated standard deviation of $29,000.

The second models the behaviour in the two larger intervals as maintaining the same ratio as the immediately previous intervals. The first three bars are identical in the two diagrams, and the shape of the tail in the second is what the distributions feels like it would look like if more detail was available. The difference in estimate of the mean is around 5%, and in the standard deviation it is about 14%.

You are not expected in this course to do anything with grouped data other than use mid-interval values in estimating mean and variance/standard deviation, but when you work with your own data you should be more cautious when basing decisions on grouped data, especially if it is skewed, than you would be using estimates based on individual data values.

Exercise 7.3

1. For the following samples of data, calculate unbiased estimates of the mean and variance of the populations the samples were drawn from:

a	32	25	29	31	27	39	24	22	31			
b	11	14	12	18	11	7	12	18	21	17	16	12
c	65	67	71	73	68	71	74	66				
d	89	78	88	79	83	84	82	95	95	77	83	
e	17	35	56	25	44	61	57	22	32	55		

Unbiased estimate of the population variance

2. For the following samples of data, calculate unbiased estimates of the mean and variance of the populations the samples were drawn from:

a)

x	12	13	14	15	16	17	18
frequency	5	9	14	11	8	3	1

b)

x	25	26	27	28	29	30	31
frequency	6	11	12	12	8	5	2

3. For the following samples of data, calculate the best estimates of the mean and variance of the populations the samples were drawn from:

a)

interval	$0 \leq x < 5$	$5 \leq x < 10$	$10 \leq x < 15$	$15 \leq x < 20$	$20 \leq x < 25$	$25 \leq x < 30$
frequency	3	9	16	14	7	2

b)

interval	$0 \leq x < 10$	$10 \leq x < 20$	$20 \leq x < 25$	$25 \leq x < 30$	$30 \leq x < 40$	$40 \leq x < 50$
frequency	83	121	87	79	106	65

4. For the following summary statistics of samples of data, calculate unbiased estimates of the mean and variance of the populations the samples were drawn from:

a) $n = 50; \sum x = 423; \sum x^2 = 4956$

b) $n = 23; \sum x = 835; \sum x^2 = 37825$

c) $n = 68; \sum x = 695; \sum x^2 = 12457$

d) $n = 8; \sum x = 657; \sum x^2 = 75688$

e) $n = 97; \sum x = -25; \sum x^2 = 154$

5. The times taken by a random sample of 15 students attending a college to travel to college were measured. The data are summarised by $\sum t = 387; \sum t^2 = 13221$. Find

a) the variance of the times in the sample

b) an unbiased estimate of the variance of the travel times taken by all students.

6. Unbiased estimates of the mean and variance of a population, based on a random sample of 10 observations, are 24.3 and 5.2 respectively. Another random observation is obtained and is measured at 21.5. Find new unbiased estimates of the mean and variance of the population, based on the full information now available.

7.4 Confidence intervals for the mean of a Normal distribution

In section 7.1 you looked informally at the rationale for an interval estimate rather than a point estimate – that it allows you to give some idea of how good the estimator is – and that is dependent on the variability of the population and of the size of the sample used.

For a Normal distribution, you know that the variance of $\overline{X}$ is $\dfrac{\sigma^2}{n}$. The standard deviation of the sampling distribution $\left(\dfrac{\sigma}{\sqrt{n}}\right)$ is known as the **standard error,** and in chapter 4 you saw that the distribution of $\overline{X}$ is also Normal,

so $\Pr\left(\mu - 1.96\dfrac{\sigma}{\sqrt{n}} < \overline{X} < \mu + 1.96\dfrac{\sigma}{\sqrt{n}}\right) = 0.95.$

The only thing is that you do not actually know where μ is, but you do know where the observed sample mean $\overline{x}$ is, and you can use a very neat piece of logic to turn this round: on 95% of occasions the sample mean will be captured by an interval centred on μ and extending $\pm 1.96\dfrac{\sigma}{\sqrt{n}}$; every time $\overline{x}$ lies in that interval, μ has to be captured by an interval of the same size centred on $\overline{x}$.

> $\overline{x} \pm 1.96\dfrac{\sigma}{\sqrt{n}}$ is the 95% **confidence interval** (CI) for the mean of a Normal distribution based on a random sample of size n.

The diagram below tries to illuminate this visually:

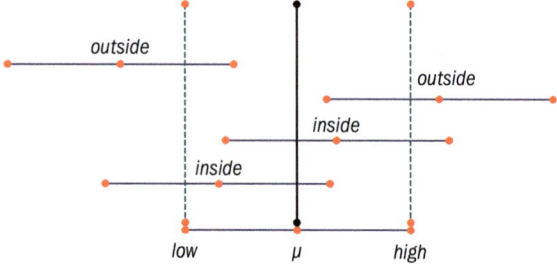

The bottom of the diagram shows $\mu - 1.96\dfrac{\sigma}{\sqrt{n}}$, μ and $\mu + 1.96\dfrac{\sigma}{\sqrt{n}}$ as the interval centred on the true population mean μ. There are then 4 intervals shown, of the same width, centred on four possible observed values of $\overline{x}$. For the bottom two, the value of $\overline{x}$ lies inside the interval centred on μ but for the top two the value of $\overline{x}$ lies outside. The bold black line shows where the value of μ is – and it cuts the two bottom intervals but not the top two intervals.

So 95% of the time the true population mean will be captured by a confidence interval constructed in this way. You should talk about the true population mean lying in the confidence interval on 95% of occasions a confidence interval is constructed rather than talking about a probability – because the population mean is not a random variable.

The usual size of confidence interval is 95%, but it is not the only possible one. It is commonly used because of the shape of the Normal distribution – the curve really starts to tail off very quickly at around 2 standard deviations from the mean.

Technically, you can construct any size of confidence interval you choose to for a continuous distribution but in practical terms there are three others which are the most common alternatives: 90%, 98% and 99%.

> For the mean of a Normal distribution based on a random sample of size n,
>
> $\bar{x} \pm 1.645 \dfrac{\sigma}{\sqrt{n}}$ is the 90% confidence interval
>
> $\bar{x} \pm 2.326 \dfrac{\sigma}{\sqrt{n}}$ is the 98% confidence interval
>
> $\bar{x} \pm 2.576 \dfrac{\sigma}{\sqrt{n}}$ is the 99% confidence interval

If you have a 95% confidence interval the tails each contain $2\frac{1}{2}\%$.

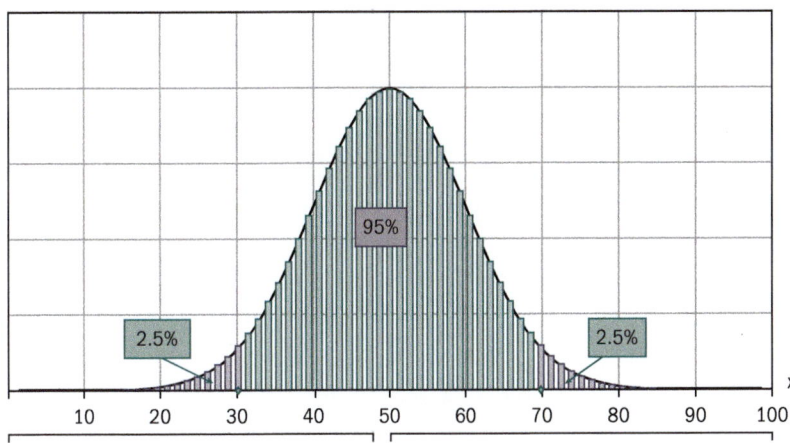

More generally if you have an $\alpha\%$ confidence interval the tails each contain $\dfrac{(100 - \alpha)}{2}\%$.

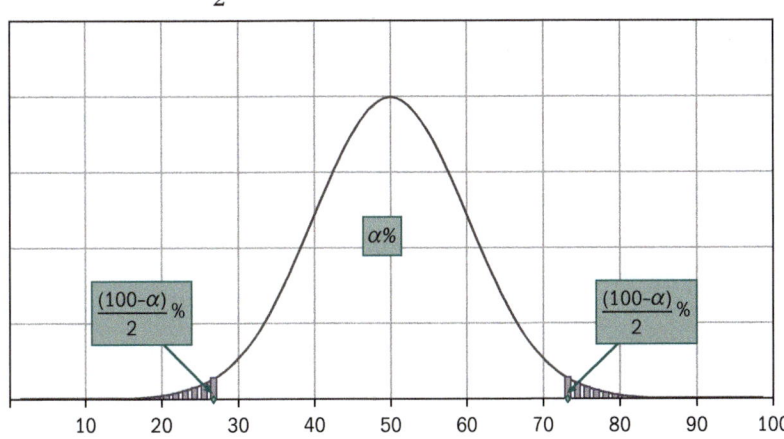

One thing to note is that in a lot of production processes, the variability of the output is quite stable – it is a function of the quality of the process, but the mean can usually be altered by a setting on the machinery, and can often move while the process is underway. So where you usually need to calculate the standard deviation (or variance) in data, in a production process you will often know the standard deviation and can use this in constructing an interval estimate for the mean.

Example 4

The weights, in grams, of bags of cocoa beans are known to have a standard deviation of 6 grams. A random sample of bags is weighed with the following results:

758, 748, 749, 752, 757, 760, 751, 745, 759, 761

Calculate a 95% confidence interval for the mean weight of a bag of cocoa beans.

$\sum x = 7540$; $\bar{x} = \frac{7540}{10} = 754$, so the 95% confidence interval is

$\bar{x} \pm 1.96\frac{\sigma}{\sqrt{n}} = 754 \pm 1.96 \times \frac{6}{\sqrt{10}} = 754 \pm 3.7 = (750.3, 757.7)$

Exercise 7.4

1. The weights, in grams, of bags of onions are known to be normally distributed and have a standard deviation of 30 grams. A random sample of bags is weighed with the following results:

    742, 775, 712, 765, 712, 703, 727, 772, 731, 781

    Calculate a 95% confidence interval for the mean weight of a bag of onions

2. The volumes, in millilitres, of bottles of water are known to be normally distributed and have a standard deviation of 9.3 millilitres. A random sample of bottles are measured with the following results (in millilitres):

    605, 617, 592, 603, 621, 599, 612, 603, 609, 614, 595, 610

    i)   Calculate a 95% confidence interval for the mean volume of a bottle of water.

    ii) Will a 90% confidence interval for the mean volume be wider or narrower than the 95% CI in part (i)?

3. The heights, in centimetres, of a type of plant are known to be normally distributed and have a standard deviation of 8 centimetres. A random sample of plants is weighed with the following results:

    35, 38, 42, 31, 33, 45, 34, 46, 40, 36

    Calculate a 90% confidence interval for the mean height of the plants.

4. The weights, in grams, of bowls made in a particular factory are known to be normally distributed and have a standard deviation of 21 grams. A random sample of bowls is weighed with the following results:

    195, 232, 211, 242, 201, 223, 247, 161, 226, 237

    i)   Calculate a 95% confidence interval for the mean weight of the bowls.

    ii) Calculate a 98% confidence interval for the mean weight of the bowls.

5. The heights, in centimetres, of a type of plant are known to be normally distributed and have a standard deviation of σ centimetres. A random sample of 25 plants is weighed and the 95% confidence interval obtained is (43.2, 47.8).

   i) what was the sample mean, $\bar{x}_{25}$?

   ii) what is the value of σ, correct to 1 decimal place?

7.5 Confidence intervals for the mean of a large sample from any distribution

In chapter 6 you met the Central Limit Theorem which said that for large samples (as a rule of thumb, for $n > 30$) the distribution of the sample means is approximately Normal. In chapter 3 you met that the variance of $\bar{X}$ is always $\dfrac{\sigma^2}{n}$, and in this chapter you have learnt that s^2 is an unbiased estimator of the population variance (and that even where the data is given in a frequency table for interval, where it is not unbiased, s^2 is still the best estimator available).

Using all of these, we can deduce that:

> For a large random sample ($n > 30$) from any population, a 95% confidence interval is given by $\left(\bar{x} - 1.96\dfrac{\sigma}{\sqrt{n}}, \bar{x} + 1.96\dfrac{\sigma}{\sqrt{n}} \right)$ when the variance is known (quite rare) or by $\left(\bar{x} - 1.96\dfrac{s}{\sqrt{n}}, \bar{x} + 1.96\dfrac{s}{\sqrt{n}} \right)$ when the population variance has had to be estimated from the sample using $s^2 = \dfrac{1}{n-1}\sum(x_i - \bar{x})^2$ or one of the computational equivalents.
>
> Other size confidence intervals are obtained by replacing 1.96 in this expression by the corresponding z score for the required level of confidence.

Note that strictly speaking this is an approximate confidence interval because you are using the Central Limit Theorem to say that the distribution of the sample means is approximately Normal.

Example 5

The weights, x, in grams, of a random sample of 50 adult hamsters are recorded. The following summary statistics were calculated:

$$\sum x = 10\,003.3, \quad \sum x^2 = 2\,002\,665$$

Calculate a 95% confidence interval for the mean weight of the hamsters.

▶ Continued on the next page

$\sum x = 10\,003.3$ $\bar{x} = \dfrac{10\,003.3}{50} = 200.07$

$s^2 = s^2 = \dfrac{1}{n-1}\left(\sum x^2 - \dfrac{\left(\sum x\right)^2}{n}\right)$

$\qquad = \dfrac{1}{49}\left(2\,002\,665 - \dfrac{10\,003.3^2}{50}\right) = 27.44..$

$\Rightarrow s = 5.24$

so the 95% confidence interval is

$\bar{x} \pm 1.96\dfrac{s}{\sqrt{n}} = 200.07 \pm 1.96 \times \dfrac{5.24}{\sqrt{50}} = 200.07 \pm 1.45 = (198.6, 201.5)$

Although the distribution of the weights of the hamsters is not known, the sample size is large enough to rely on the Central Limit Theorem for the distribution of the sample means.

It is sensible to keep extra accuracy during the working out, and only round at the end, but don't give too much accuracy at the end as it implies a level of knowledge which is not justified. The width of the interval is really the thing to consider when deciding the appropriate accuracy – here $\pm 1.96\dfrac{s}{\sqrt{n}}$ was ± 1.45 and the CI limits were calculated using both the mean and width to 2 decimal places before rounding to one decimal place in the final statement of the CI. Even though there are 4 signifiant figures given in the CI limits – the accuracy is really only 2 significant figures in the interval width.

Example 6

A machine is supposed to produce metal rods which are 5.7 cm long.
A random sample of 100 rods produced by the machine are measured and the lengths, x cm, are summarised below.

x	5.60 – 5.65	5.65 – 5.70	5.70 – 5.75	5.75 – 5.80
frequency	15	31	36	18

Calculate a 95% confidence interval for the mean length of a rod produced by the machine.

The mid interval values, m, are 5.625, 5.675, 5.725 and 5.775

$\sum mf = 570.35$ $\bar{x} = \dfrac{570.35}{100} = 5.7035$ $\sum m^2 f = 3253.218$

$s^2 = s^2 = \dfrac{1}{n-1}\left(\sum m^2 f - \dfrac{\left(\sum mf\right)^2}{\sum f}\right)$

$\qquad = \dfrac{1}{99}\left(3253.218 - \dfrac{570.35^2}{100}\right) = 0.002286$

$\Rightarrow s = 0.047808$

so the 95% confidence interval is

$\bar{x} \pm 1.96\dfrac{s}{\sqrt{n}} = 5.7035 \pm 1.96 \times \dfrac{0.047808}{\sqrt{100}} = 5.7035 \pm 0.00937 = (5.694, 5.713)$

Here, you do not have the exact values of the lengths but the data is grouped in equal size intervals.

There is an extra layer of uncertainty involved in this situation, so if there is any doubt about the appropriate level of accuracy, you should be inclined to choose less accuracy rather than more.

Exercise 7.5

1. The summary statistics of a number of random samples are given below.
   For each one, give

   i) a 95% confidence interval for the mean
   ii) a 90% confidence interval for the mean

   a) $n = 50;\ \sum t = 357;\ \sum t^2 = 12721$

   b) $n = 37;\ \sum x = 1635;\ \sum x^2 = 75325$

   c) $n = 62;\ \sum v = 651;\ \sum v^2 = 12712$

   d) $n = 71;\ \sum w = 617;\ \sum w^2 = 27688$

   e) $n = 97;\ \sum p = -32.4;\ \sum p^2 = 174$

2. In Q1, does it make any difference if the populations are known to
   be normally distributed?

3. A random sample of 45 cans of mixed beans gave unbiased estimates of 324.6
   grams and 6.2 grams^2 for the population mean and variance respectively.
   Calculate a 95% confidence interval for the population mean weight.

4. The times taken to complete a task were recorded for a random sample
   of 60 workers. The sample mean was 28.3 minutes and the unbiased estimate
   of the population variance was 31.2 minutes^2. Find a 90% confidence interval
   for the mean time taken to complete the task.

5. A random sample of 75 cans of paint gave unbiased estimates of
   503.2 ml and 14.3 ml^2 for the population mean and variance respectively.
   Calculate a 95% confidence interval for the population mean volume.

6. A random sample of 40 bottles of cleaning fluid gave unbiased estimates
   of 102.1 cl and 7.3 cl^2 for the population mean and variance respectively.
   Calculate a 98% confidence interval for the population mean volume.

7.6 Confidence intervals for a proportion

In chapter 9 of S1 you met the normal approximation to the Binomial distribution.
- The Binomial distribution $B(n, p)$ can be approximated by the Normal
  distribution $N(np, npq)$ provided both np and $nq = n(1 - p)$ are greater than 5.
- The continuity correction must be used because a discrete distribution
  is being approximated by a continuous distribution.

When you met the Binomial distribution first, the condition that it required
only two outcomes may have seemed to be so constraining that the Binomial
would have little practical use, until it became clear that the outcomes for any

situation could be partitioned into two categories (often categorized as 'success' for those you would count, and 'failure' for those that you would not) such as 'throw a six' with a fair die, or 'the height of a plant is over 27 cm'. The Binomial distribution is then giving you the number of times, x, that some condition is seen out of n observations of the population – which gives the proportion of times the condition is seen $\left(= \frac{x}{n}\right)$ to be calculated.

Imagine you wanted to estimate the proportion of people who are left handed. If you took a random sample of 100 people and found 20 were left handed, your best estimate of the proportion in the population would be 20%. If you had taken a random sample of 500 and had found 100 were left handed then your (point) estimate would again be 20%, but it is based on much more evidence than was available in the first case and you would feel that the precision of your estimate should be rather better – and any interval estimate should be narrower for the large sample than for the smaller one.

You need to ensure that the usual conditions for using the Binomial are met, so there needs to be a fixed number of trials, only two possible outcomes, with a constant probability of 'success', and that the trials are independent of one another.

Using all of these, we can deduce that:

For a large random sample, size n, a 95% confidence interval for the proportion, p, with a particular property is given by

$$\left(p_s - 1.96\sqrt{\frac{p_s q_s}{n}}, \, p_s + 1.96\sqrt{\frac{p_s q_s}{n}}\right)$$ where p_s is the sample proportion with the

property, and $q_s = 1 - p_s$.

Other size confidence intervals are obtained by replacing 1.96 in this expression by the corresponding z score for the required level of confidence.

Note that strictly speaking this is an approximate confidence interval because you are using the Central Limit Theorem to say that the distribution of the sample proportion is approximately Normal, but it has a little more fuzziness as well – you have not applied the continuity correction and not adjusted to get an unbiased estimate of the variance. However, the extra precision which is gained by including those improvements is small and the effort required to make the adjustments needed to include them is large enough that it is usual not to do so – and you are not expected to in this course.

Example 7

A large urban area is considering introducing making two of the five lanes into high-occupancy vehicle lanes on one of the main routes into the area at peak travel times to try to ease traffic congestion. Planning officials commission a study to estimate the proportion of vehicles currently carrying only the driver.

The study recorded 653 vehicles entering the area on that route between 0810 and 0815 on a randomly selected ordinary working day. Analysis of the video recording suggested that 362 of these vehicles carried only the driver.

i) Calculate a 95% confidence interval for the proportion of vehicles currently carrying only the driver on that route at peak times.

ii) If the planning officials want to collect more data, suggest how they might do it in order to get the best information possible for their situation.

> Part **ii** is beyond what you would be asked in the examination.

i) From the data, $p_s = \dfrac{362}{653}$ $(= 55.4\%)$, $n = 653$, $q_s = \dfrac{291}{653}$ so the 95% confidence interval is given by

$$\left(p_s - 1.96\sqrt{\dfrac{p_s q_s}{n}},\ p_s + 1.96\sqrt{\dfrac{p_s q_s}{n}} \right) = \left(\dfrac{362}{653} - 1.96\sqrt{\dfrac{362 \times 291}{653^3}},\ \dfrac{362}{653} + 1.96\sqrt{\dfrac{362 \times 291}{653^3}} \right) = (0.516, 0.592)$$

ii) The expense of collecting data on this situation comes in two parts – video recording the traffic, and then analyzing the video to identify vehicles with only the driver. Collecting only one block of data probably keeps those costs to a minimum, but recording five one minute blocks would cost the same for the analysis, and allow them to see if there is any variation in proportion at different times. Most of the costs to do the recording would be to get the people and equipment there and set up – having them stay for an hour longer to record a number of short blocks would cost very little more.

Exercise 7.6

1. The sample size, n, and number of people, s, in the sample satisfying a particular criterion is given below for a number of surveys. Assuming these are random samples, calculate give

   i) a 95% confidence interval for the population proportion

   ii) a 90% confidence interval for the population proportion satisfying the criterion in each survey.

      a) $n = 120$; $s = 37$

      b) $n = 85$; $s = 12$

      c) $n = 62$; $s = 25$

      d) $n = 71$; $s = 22$

      e) $n = 97$; $s = 14$

2. A random sample of 100 bolts was measured and 12 of them were found to lie outside the production tolerance limits. Find a 95% confidence interval for the proportion of bolts which lie outside the production tolerance levels.

3. A random sample of 80 city workers is taken and 23 of them usually travel to work by bus. Find a 95% confidence interval for the proportion of city workers who usually travel to work by bus.

4. During the morning 300 cars were observed on a busy road and 143 of them carried no passengers.

   i) Find a 90% confidence interval for the proportion of cars on the road which carry no passengers.

   ii) state any assumptions that you have had to make in constructing the confidence interval.

   iii) an α confidence interval is constructed using the same sample. The interval has width 0.15. Find the value of α.

5. A random sample of 250 items for which the prices were compared in two large shop found that Easipay were cheaper on 112 items than Valustore.

   i) Find a 95% confidence interval for the proportion of items for which Easipay are cheaper than Valustore.

   ii) Estimate the size of the sample needed for an approximate 95% confidence interval to have width 0.05.

6. A random sample of n people found that 53 of them preferred to listen to news on the radio than in any other format. A symmetric confidence interval for the population proportion who prefer to listen to news on the radio is $0.167 < p < 0.271$.

   i) Find the midpoint of this interval and use it to calculate the size of the sample.

   ii) Find the confidence level of this interval.

Summary exercise 7

1. For the following sample of data, calculate unbiased estimates of the mean and variance of the population:

   42 25 27 53 41 62 27 35 56 37 46 51 29

2. For the following sample of data, calculate unbiased estimates of the mean and variance of the population:

x		22	23	24	25	26	27	28
frequency		17	23	29	45	32	19	8

3. For the following sample of data, calculate the best estimates of the mean and variance of the population:

interval	$0 \leq x < 5$	$5 \leq x < 10$	$10 \leq x < 15$	$15 \leq x < 20$	$20 \leq x < 25$	$25 \leq x < 30$
frequency	19	37	55	67	26	8

4. For the following summary statistics of samples of data, calculate unbiased estimates of the mean and variance of the populations the samples were drawn from:

a) $n = 90; \sum x = 353; \sum x^2 = 4172$

b) $n = 19; \sum x = 689; \sum x^2 = 54369$

5. The heights, in centimetres, of a type of plant are known to be normally distributed and have a standard deviation of 11 centimetres. A random sample of plants is weighed with the following results:

75, 68, 86, 92, 71, 83, 92, 65, 78, 82

Calculate a 90% confidence interval for the mean height of the plants.

6. The heights, in centimetres, of a type of plant are known to be normally distributed and have a standard deviation of σ centimetres. A random sample of 35 plants is weighed and the 95% confidence interval obtained is (36.7, 44.9).

i) what was the sample mean, $\bar{x}_{35}$?

ii) what is the value of σ, correct to 1 decimal place?

7. The summary statistics of two random samples are given below. For each one, give

i) a 95% confidence interval for the mean

ii) a 90% confidence interval for the mean

a) $n = 130; \sum t = -67.5; \sum t^2 = 12231$

b) $n = 82; \sum x = 952; \sum x^2 = 14526$

8. The times taken to complete a puzzle were recorded for a random sample of 70 students. The sample mean was 12.3 minutes and the unbiased estimate of the population variance was 9.2 minutes². Find a 90% confidence interval for the mean time taken to complete the puzzle.

9. A random sample of 75 tiles was measured and 8 of them were found to have slight flaws. Find a 95% confidence interval for the proportion of tiles with slight flaws.

10. A random sample of 120 commuters is taken and 28 of them say they would like to change job in order to cut down travel time. Find a 95% confidence interval for the proportion of commuters would like to change job in order to cut down travel time.

11. A random sample of n people found that 35 of them did not vote in a recent referendum. A symmetric confidence interval for the population proportion who prefer to listen to news on the radio is $0.187 < p < 0.383$.

i) Find the midpoint of this interval and use it to calculate the size of the sample.

ii) Find the confidence level of this interval.

EXAM-STYLE QUESTIONS

12. Diameters of table tennis balls are known to be normally distributed with mean μ and standard deviation σ. A random sample of 125 table tennis balls was taken and the diameters, d cm, were measured. The results are summarised by $\sum d = 498.2; \sum d^2 = 1985.78$.

i) Calculate unbiased estimates of μ and σ^2

ii) Calculate a 96% confidence interval for μ

iii) 200 random samples of 125 balls are taken and a 96% confidence interval is calculated for each sample. How many of these interval would you expect **not** to contain μ?

13. A survey was conducted to find the proportion of people who attend an aerobics class on a regular basis. It was found that 89 people out of a random sample of 190 people attend an aerobics class on a regular basis.

i) Calculate a 98% confidence interval for the true proportion of people who attend an aerobics class on a regular basis.

A second survey to find the proportion of people who attend an aerobics class on a regular basis was conducted in a shopping centre on a Wednesday afternoon.

ii) Give a reason why this is not a satisfactory sample.

14. The weights, in grams, of packets of flour are distributed with mean μ and standard deviation 15.3. A random sample of 85 packets is taken. The mean weight of the sample is found to be 748 g. Calculate a 97% confidence interval for μ.

15. a) Give a reason why sampling would be required in order to reach a conclusion about

   i) The mean height of adult males in England, [1]

   ii) The mean weight that can be supported by a single cable of a certain type without the cable breaking. [1]

   b) The weights, in kg, of sacks of potatoes are represented by the random variable X with mean μ and standard deviation σ. The weights of a random sample of 500 sacks of potatoes are found and the results are summarised below.

   $n = 500, \sum x = 9850, \sum x^2 = 194\,125.$

   i) Calculate unbiased estimates of μ and σ^2. [3]

   ii) A further random sample of 60 sacks of potatoes is taken. Using your values from part b) i), find the probability that the mean weight of this sample exceeds 19.73 kg. [4]

   iii) Explain whether it was necessary to use the Central Limit Theorem in your calculation in part b) ii). [2]

Cambridge International A Level and AS Mathematics 9709 Paper 71 Q7 November 2010

16. In a survey of 1000 randomly chosen adults, 605 said that they used email. Calculate a 90% confidence interval for the proportion of adults in the whole population who use email. [3]

Cambridge International A Level and AS Mathematics 9709 Paper 71 Q1 November 2010

17. The lengths of sewing needles in travel sewing kits are distributed normally with mean μ mm and standard deviation 1.5 mm. A random sample of n needles is taken. Find the smallest value of n such that the width of a 95% confidence interval for the population mean is at most 1 mm. [4]

Cambridge International A Level and AS Mathematics 9709 Paper 71 Q2 November 2009

18. Heights of a certain species of animal are known to be normally distributed with standard deviation 0.17 m. A conservationist wishes to obtain a 99% confidence interval for the population mean, with total width less than 0.2 m. Find the smallest sample size required. [4]

Cambridge International A Level and AS Mathematics 9709 Paper 71 Q2 November 2013

19. The daily takings, $x, for a shop were noted on 30 randomly chosen days. The takings are summarised by $\sum x = 31\,500,$ $\sum x^2 = 33\,141\,816.$

   i) Calculate unbiased estimates of the population mean and variance of the shop's daily takings. [3]

   ii) Calculate a 98% confidence interval for the mean daily takings. [3]

The mean daily takings for a random sample of n days is found.

iii) Estimate the value of n for which it is approximately 95% certain that the sample mean does not differ from the population mean by more than $6.   [3]

Cambridge International A Level and AS Mathematics 9709 Paper 7 Q6 May/June 2007

20. a) The time taken by a worker to complete a task was recorded for a random sample of 50 workers. The sample mean was 41.2 minutes and an unbiased estimate of the population variance was 32.6 minutes². Find a 95% confidence interval for the mean time taken to complete the task. [3]

b) The probability that an α% confidence interval includes only values that are lower than the population mean is $\frac{1}{16}$. Find the value of α.   [2]

Cambridge International A Level and AS Mathematics 9709 Paper 71 Q2 May/June 2011

Chapter summary

- A point estimate, such as $\bar{x}$, gives the best single value estimate of the location of a parameter.
- An interval estimate gives a range of values designed to give both an estimate of the location of the parameter and some indication of how precise or reliable that estimate is.
- $\bar{x} = \frac{1}{n}\sum_{i=1}^{n} X_i$ is the best unbiased estimate of the population mean μ
- $s^2 = \frac{1}{n-1}\sum(x_i - \bar{x})^2 = \frac{1}{n-1}\left(\sum x^2 - n\bar{x}^2\right) = \frac{1}{n-1}\left(\sum x^2 - \frac{(\sum x)^2}{n}\right)$, is an unbiased estimator of the population variance, σ^2.
- $\bar{x} \pm 1.96\frac{\sigma}{\sqrt{n}}$ is the 95% confidence interval for the mean of a Normal distribution based on a random sample of size n.
- $\bar{x} \pm 1.96\frac{s}{\sqrt{n}}$ is the 95% confidence interval for the mean, μ, of a distribution based on a large random sample of size $n(n > 30)$.

8 Hypothesis testing for discrete distributions

Psychology, medicine, archaeology, manufacturing processes and many other areas of modern life routinely use statistical testing to make decisions about what constitutes unusual behaviour or indicates a departure from the norm in a group. The consequences of making wrong decisions may vary from minor inconvenience, to financial losses in millions of pounds, to life and death, so understanding the logic of testing is critical for people working in these areas.

Objectives

After studying this chapter you should be able to:

- Understand the nature of a hypothesis test, the difference between one-tail and two-tail tests, and the terms null hypothesis, alternative hypothesis, significance level, rejection region (or critical region), acceptance region and test statistic;

- Formulate hypotheses and carry out a hypothesis test in the context of a single observation from a population which has a binomial or Poission distribution, using either direct evaluation of probabilities or a normal approximation, as appropriate;

- Understand the terms Type I error and Type II error in relation to hypothesis tests;

- Calculate the probabilities of making Type I and Type II errors in specific situations involving tests based on a normal distribution or direct evaluation of binomial or poisson probabilities.

Before you start

You should know how to:

1. Calculate probabilities for the Binomial distribution. If $X \sim B(10, 0.3)$ find the largest value of x for which $P(X > x) < 0.05$
$P(X = 0) = 0.7^{10} = 0.028$;
$P(X = 1) = 10 \times 0.3 \times 0.7^9 = 0.121$
so $x = 0$ is the largest

2. Calculate probabilities for the Poisson distribution. If $X \sim P(7)$ find $P(X < 3)$

$$P(X < 3) = e^{-7}\left(1 + 7 + \frac{7^2}{2}\right) = 0.0296$$

Skills check:

1. If $X \sim B(15, 0.1)$ find the least value of x for which $P(X > x) < 0.1$

2. If $X \sim P(10)$ find the largest value of x for which $P(X < x) < 0.025$

8.1 The logical basis for hypothesis testing

Anil tossed a coin five times and got only 1 head.
He thought the coin was biased.

Vanya tossed another coin 500 times and got 129 heads.
She thought her coin was biased.

Could you make a convincing argument in either case that the coin was biased? Or that the coin was not biased?

You could repeat their experiments with a fair coin and see how often you got something like their observations.
If it rarely happens then you might be inclined to agree the coin was biased.
You might be able to do this practically in Anil's case but in Vanya's it would take a very long time.

You could use an electronic simulation, or compare their results against calculated probabilities (Vanya's would need the Normal approximation).

A critical principle is to look for the likelihood of seeing at least as extreme a result as the observed.
Cases where none of the 5 tosses give a head would be taken as well as those which give 1 head.

The second critical principle is to also include the cases of 0 or 1 tail (5 or 4 heads) as being 'at least as extreme' **unless** there was some reason to believe before Anil tossed the coin that the coin would show more heads than tails if it was biased.

It is easier to see situations where it might apply if you were looking at, say, the proportion of faulty items on a production line. You would only be interested in whether there had been an increase in the proportion.

'As bad or worse' is another way to express the idea of 'at least as extreme a result'.

For Anil's case you can work out the probability distribution exactly:

Number of heads	0	1	2	3	4	5
Probability	0.03125	0.15625	0.3125	0.3125	0.15625	0.03125

The likelihood of getting 0, 1, 4 or 5 heads is 0.375 or $\frac{3}{8}$.
This is less than half, but more than one third. You would not refer to something which happens more often than 1 in 3 times as a 'rare event'.

Similarly, it would not be reasonable to try to argue that this outcome provided positive support for the proposition that the coin **was fair**.
If you tossed a coin which only came up heads 30% of the time then 1 head in 5 tosses would be very common.

Be very careful to say explicitly that there is not sufficient evidence to say the coin is biased. Never say that it provides evidence that the coin is unbiased.

The best you can say is that Anil's result does not provide convincing evidence that the coin is biased.

Vanya's observation in her experiment actually has a higher proportion of heads than Anil's.
However, with a large number of trials, the proportion seen is much more likely to be close to the probability and the likelihood of seeing Vanya's result with a fair coin is almost vanishingly small.

What happens if the situation is not as clear cut as either of these?

Generally, the **null hypothesis H_0** is the probability basis under which you consider a test.

For the coin tossing example the null hypothesis would be $H_0 : p = \frac{1}{2}$.

The **alternative hypothesis $H_1 : p \neq \frac{1}{2}$** is what you would turn to if you felt there was convincing evidence that H_0 was not true.

The principle of hypothesis testing is to compare where the value of an observed statistic (**the test statistic**) lies within the sampling distribution under the null hypothesis.

If it would be classed as a 'rare event' (and the **significance level of a test** is whatever level we use to judge 'rare') then you look for an alternative explanation.

If the event observed is not an unusual occurrence when the null hypothesis is true, it will also not be unusual for a lot of other situations, so accepting the null hypothesis is a very, very weak statement; there is no positive supporting evidence that it is true.

In many situations, you can determine the significance level of a test exactly (= the proportion to be classed as 'rare').

5% is the standard level.

If you decrease the significance level, then you are less likely to reject the null hypothesis incorrectly (i.e. when it is true) - but you will automatically be more likely to accept the null hypothesis when it is not true.

The 5% standard level is used to give the best trade-off between the 2 types of error. If it is more important that you don't reject H_0 wrongly you can use a 1% significance level.
Where it is more important to not accept H_0 wrongly you can use a significance level above 5%.

In this chapter you will only be testing a discrete distribution (the Binomial or Poisson) and in chapter 9 you will look at testing the mean of a distribution using the Normal.

For discrete cases only taking integer values, the decision rule may be in the form 'Reject H_0 if $X \le 1$, or if $X \le 2$ …'. Testing with $n = 22$, for $p = 0.25$ against $p < 0.25$ the probability of these events happening if the null hypothesis is true are 1.5% and 6.1%.

In this case you cannot have a 5% test – you could have a 1.5% test or a 6.1% test.

For this examination the size of the test stated is the maximum size which can be used – so the decision rule here will be 'Reject H_0 if $X \le 1$' and the size of the test will be 1.5%

> An alternative approach would be to choose the size closest to your target level.

To construct a test (critical region approach):

- Identify formally the null and alternative hypotheses.
  - The null hypothesis must give a distribution for the test statistic
    In this context it is always in the form
    $H_0: p =$ or $H_0 : \lambda =$
  - The alternative hypothesis is in one of the forms:
    $H_1: p >$ or $p <$ or $p \ne$ (or similar for the Poisson)
- Give an explicit statement of the decision rule to be used.
  (Including the significance level you are using)

- Apply the decision rule to the observed value of the test statistic.
  Avoid categorical statements in your conclusion statements.
  A test allows you to state either
  - Accept H_0.
    This is **not because the evidence supports it**, and you should not say this. You can only conclude there is not significant evidence against H_0.
  - Reject H_0.
    Conclude that there is evidence, significant at your chosen significance level, that the mean or probability is not — (or of an increase/decrease in mean or probability if it is one tailed).
- Give the conclusions of the test in the context of the problem.

If you are told that the sample was discovered not to be random, then the conclusion of your test is invalidated.

Example 1

The proportion of applications from women to a particular course has been stable at 35% over a number of years. The course is revised.
A random sample of 50 applications is taken to see if the proportion of women applying for the new course has changed.
a) Give the null and alternative hypotheses you would use in the test.
b) In fact 31 out of the 50 applications in the above test were from women, and the null hypothesis was rejected at the 5% level of significance. State the conclusion in context.

a) $H_0: p = 0.35$ *vs* $H_1: p \neq 0.35$ - You are only looking at whether the proportion has changed – the revision might make it more or less attractive to women.
b) Reject H_0 and conclude there is sufficient evidence at the 5% level to suggest that the proportion of applications from women is not 35%.

Example 2

The number of motorbikes going over the speed limit on a road in a housing estate in a ten minute period between 4 p.m. and 7 p.m. may be modelled by a Poisson random variable with parameter 2.2.
Traffic humps are installed on the road as a traffic calming measure.
a) Give the null and alternative hypotheses you would use in a test to see if the number of motorbikes breaking the speed limit on the road had been reduced.
b) In a randomly selected 10 minute period between 4 p.m. and 7 p.m. after the change was made there were two motorbikes breaking the speed limit, and the null hypothesis was not rejected at the 5% level of significance. State the conclusion in context.

a) $H_0: \lambda = 2.2$ *vs* $H_1: \lambda < 2.2$ - You are only looking at whether the average rate had been **reduced**.
b) Accept H_0 and conclude there is not sufficient evidence at the 5% level to suggest that the average number of motorbikes breaking the speed limit in a 10 minute period has been reduced.

When you do a hypothesis test, there are two possible types of error.

Type 1 error:
The null hypothesis is actually true and you reject it.
This is the **significance level of the test**.

We will return to this later in the chapter and deal with it in more detail.

Type II error:
You accept the null hypothesis and it was not true.
The likelihood of this happening depends on what the true value of the parameter is.

Exercise 8.1

1. The proportion of primary school teachers who have passed A-level Maths is 23%. The government launches a campaign to encourage people with A-level Maths to consider teaching in primary schools.

   A random sample of 40 applications for training is taken to see if the proportion of people with A-level Maths has increased. Give the null and alternative hypotheses you would use in the test.

2. 13 out of the 40 applications in the above test were from people with A-level Maths, and the null hypothesis was not rejected at the 5% level of significance.
   State the conclusion in context.

3. The number of accidents on a particular stretch of road can be modelled by a Poisson distribution with a mean of 0.45 accidents per day on weekdays. Give the null and alternative hypotheses you would use in a test to see if the number of accidents on that stretch of road is different at the weekend.

4. Over a month which has 8 weekend days in it there were 7 accidents recorded on that stretch of road, and the null hypothesis was rejected at the 5% level of significance.
   State the conclusion in context.

8.2 Critical region

In all cases the null hypothesis must provide a probability distribution for the test statistic.

The basis of **all** tests is to see if the observations lie in the most extreme part of this distribution - called the **critical region**, or the **rejection region**. You reject H_0 if the observations lie in this region. All other values lie in the **acceptance region** of the test.
The probability of this happening when H_0 is actually true is called the size of the test. You use 5% as the common default size.

A good test is one where the probabilities of both possible errors are as small as possible.
For a fixed sample, however, you can only reduce the probability of a Type I error by widening the net to include cases previously in the critical region. This immediately reduces the chance of rejecting H_0 when it is not true but also increases the probability of the Type II error.

A smaller significance level means that you are:

- Less likely to reject H_0 incorrectly
- More likely to be unable to reject H_0 when it is not true.

A larger significance level means that:

- You are more likely to be able to reject H_0 when it is not true but you are also more likely to reject it when it is true.

Consider the case where the null hypothesis is that $\lambda = 7.5$.
The sampling distribution looks like this.

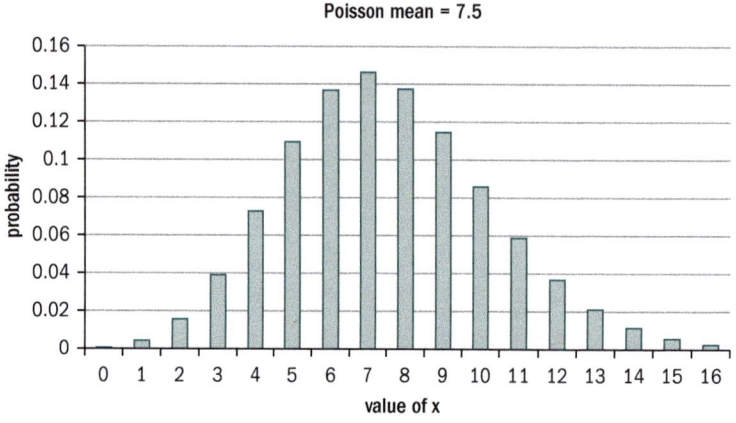

In the examination you will only be asked questions where there are a small number of values needing to be calculated in order to find the critical regions.
We will show the full distributions in some examples so you can develop a feel for what is going on.

If you wanted to test that the parameter λ was different from 7.5, then very low or very high observed values would be considered 'extreme' and would push you away from the null hypothesis.

This would be a **2-tailed** test, because values in both tails of the distribution would be considered extreme.

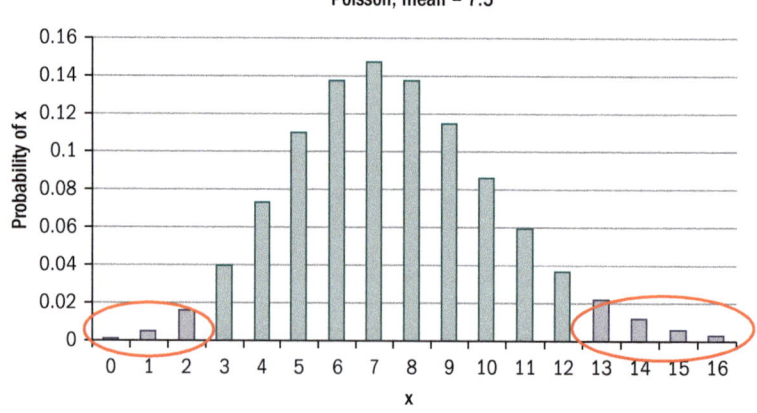

Note that there are values above 16 not shown on the graph here.

At the 5% level of significance, the 2-tailed test will reject H_0 if X is 0, 1 or 2 or if it is 13 or more.

There is a probability of 2.0% at the bottom end and another 2.2% at the top end, but this is as close to 5% as it is possible to get.

If you had some reason to expect that λ would only be lower than 7.5, then only very low observed values would be considered 'extreme'. This would be a **1-tailed test**.

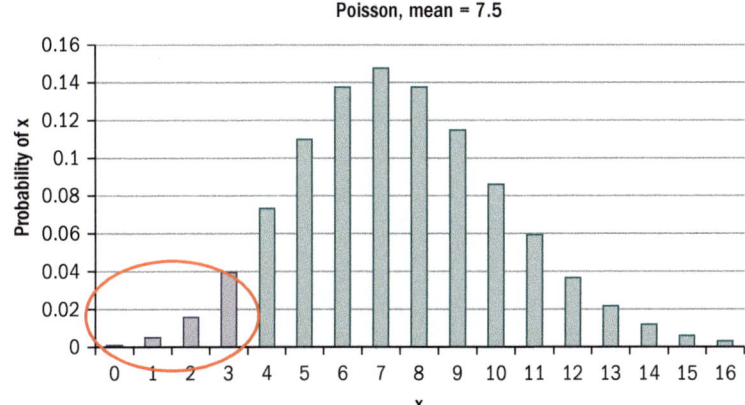

The 1-tailed test at the lower end has a critical region of X being 0, 1, 2 or 3, which has a probability of 5.9%.
If the critical region only included 0, 1 and 2 then the probability of seeing a value in the critical region when H_0 was true would be 2.0%.
There is nothing in between a 2.0% test and a 5.9% test because the Poisson is a discrete distribution. You would go for the one which is closer to 5% **unless** it was specified that the level was to be no more than 5%.

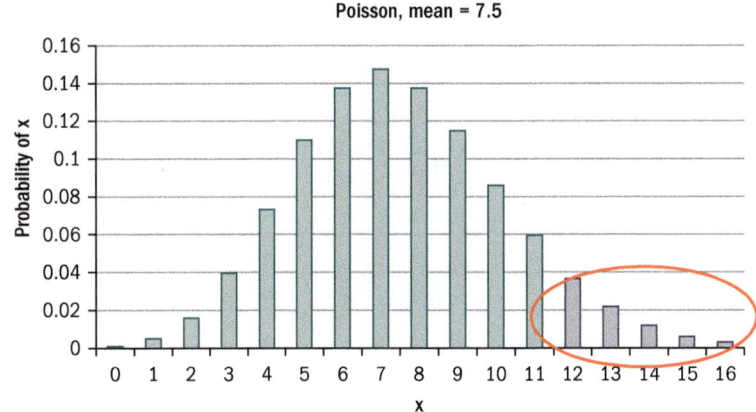

Similarly if you expected λ to only be higher than 7.5.

The 1-tailed test at the upper end has a critical region of X being 12 or more, which has a probability of 4.3%.
If the critical region included 11 as well then the probability of seeing a value in the critical region when H_0 was true would be 7.9%.

Consider the B(15, 0.5) distribution.
The table shows both the individual probabilities and the cumulative probabilities, and the graph shows the shape of the distribution.

x	$P(X = x)$	$P(X \leq x)$
0	0.00003	0.00003
1	0.00046	0.00049
2	0.00320	0.00369
3	0.01389	0.01758
4	0.04166	0.05923
5	0.09164	0.15088
6	0.15274	0.30362
7	0.19638	0.50000
8	0.19638	0.69638
9	0.15274	0.84912
10	0.09164	0.94077
11	0.04166	0.98242
12	0.01389	0.99631
13	0.00320	0.99951
14	0.00046	0.99997
15	0.00003	1.00000

While the individual probabilities are helpful in understanding the shape of the distribution and getting a sense of where the extreme values will be, it is the table of cumulative values which gives you the critical region easily.

For a 5%, 2-tailed test you want to have 2.5% at each end. 0–3 have a total probability of 1.76% and 0–4 is 5.92%, so the critical region would include 0–3 at the bottom.

At the top end, you are comparing the cumulative probabilities with a target of 0.975 to cut off 2.5%, so the rest of the critical region will be 12–15, and you need to be very careful here.
The cumulative probability in the highlighted cell (for $x = 11$) is the closest to the target of 0.975, but the first value in the critical region at the top end is 12 since $P(x \geq 11) > 0.025$.

A 5% 1-tailed test for a decrease in p would have critical region containing just 0–3 (with an exact significance level of 1.8%) because 0–4 would be greater than 5%. This shows one of the inescapable difficulties with testing discrete distributions with small numbers of likely values – you may not be able to get close to the desired significance level.

However, you will not have a table of probabilities – although you could take the time to generate it for yourself. You won't be expected to generate large numbers of probabilities to identify the critical region.

This table and graph show the B(15, 0.2) distribution.

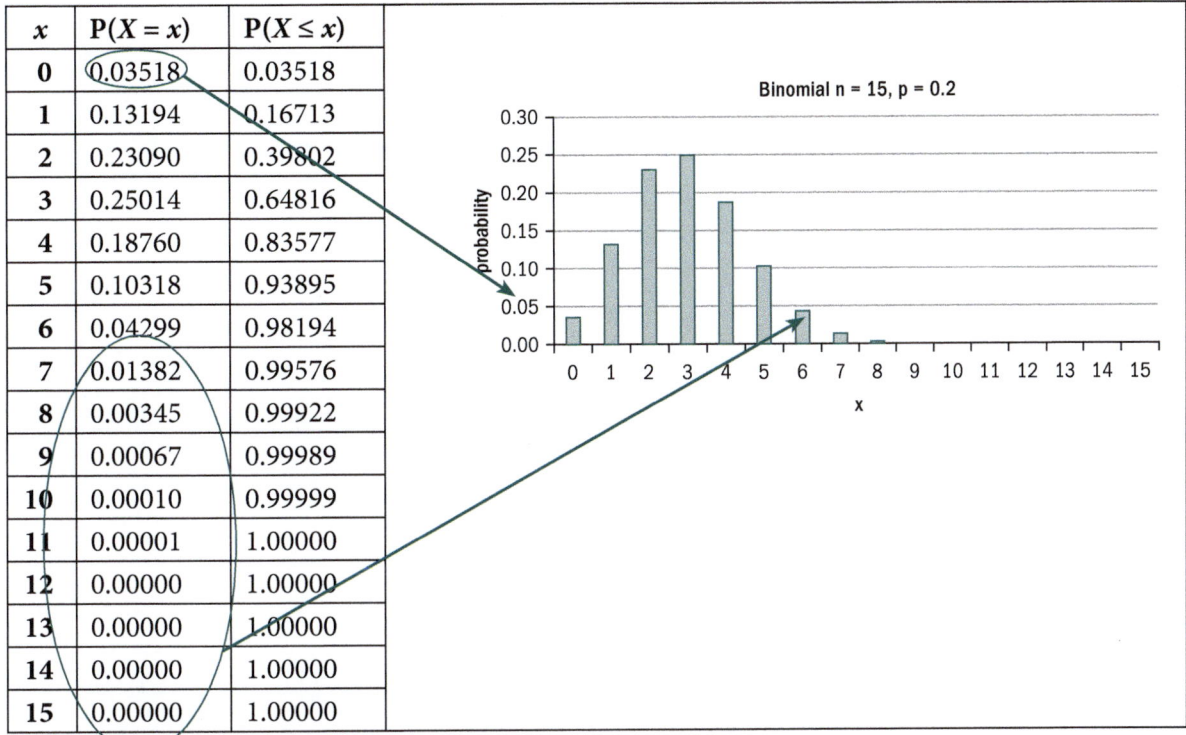

x	P(X = x)	P(X ≤ x)
0	0.03518	0.03518
1	0.13194	0.16713
2	0.23090	0.39802
3	0.25014	0.64816
4	0.18760	0.83577
5	0.10318	0.93895
6	0.04299	0.98194
7	0.01382	0.99576
8	0.00345	0.99922
9	0.00067	0.99989
10	0.00010	0.99999
11	0.00001	1.00000
12	0.00000	1.00000
13	0.00000	1.00000
14	0.00000	1.00000
15	0.00000	1.00000

The only sensible 2-tailed 5% test would have a critical region of 0 at the bottom (with probability of 3.5%) and >6 (i.e. 7 – 15) at the top (with probability 1 – 0.98194 = 1.8%). It would not make sense to not be able to reject any low values at all in a two tailed test yet the only option gives too high a probability of a Type I error. In this case you would not be able to construct a test (and you would not be asked to in the context of this course).

Remember that for Binomial distributions with large n and Poisson distributions with large values of λ you can approximate the exact distribution with a Normal distribution – which will allow tests to be carried out without burdensome calculations – you will cover this in chapter 9.

Alternative method of carrying out a test (probability approach)

You need to be familiar with critical (rejection) and acceptance regions, but there is a second approach possible in carrying out a hypothesis test. That is to work out the probability of seeing a result 'as least as unusual (far from the null hypothesis parameter value)' as the one observed. Worked examples in later sections of this chapter will show both approaches.

Exercise 8.2

For the following hypothesis tests, state whether you would use a 1 or 2 tailed test, and find the critical region and the exact significance level it will give. (Tests are 5% except in two cases highlighted in bold)

1. For a Binomial with $n = 10$, testing $H_0: p = 0.5$ against $H_1: p \neq 0.5$ with a 5% level of significance.

2. For a Binomial with $n = 10$, testing $H_0: p = 0.5$ against $H_1: p < 0.5$ with a 5% level of significance.

3. For a Binomial with $n = 10$, testing $H_0: p = 0.5$ against $H_1: p \neq 0.5$ with **a 2% level of significance**.

4. For a Binomial with $n = 12$, testing $H_0: p = 0.3$ against $H_1: p \neq 0.3$ with a 5% level of significance.

5. For a Binomial with $n = 12$, testing $H_0: p = 0.3$ against $H_1: p < 0.3$ with a 5% level of significance.

6. For a Binomial with $n = 15$, testing $H_0: p = 0.8$ against $H_1: p \neq 0.8$ with a 5% level of significance.

7. For a Binomial with $n = 15$, testing $H_0: p = 0.8$ against $H_1: p > 0.8$ with a 5% level of significance.

8. For a Poisson, testing $H_0: \lambda = 2$ against $H_1: \lambda < 2$ with a 5% level of significance.

9. For a Poisson, testing $H_0: \lambda = 2$ against $H_1: \lambda \neq 2$ with a 5% level of significance.

10. For a Poisson, testing $H_0: \lambda = 8.5$ against $H_1: \lambda < 8.5$ with a 5% level of significance.

11. For a Poisson, testing $H_0: \lambda = 8.5$ against $H_1: \lambda > 8.5$ with a 5% level of significance.

12. For a Poisson, testing $H_0: \lambda = 8.5$ against $H_1: \lambda \neq 8.5$ with a **10% level of significance**.

8.3 Type I and Type II errors

In conducting a hypothesis test you end up deciding either to accept the null hypothesis or to reject it. In either case it is possible to be incorrect – giving rise to the two types of error identified at the end of section 8.1.

> **Type 1 error:**
> The null hypothesis is actually true and you reject it.
> This is the **significance level of the test**.
> **Type II error:**
> You accept the null hypothesis and it was not true.
> The likelihood of this happening depends on what the true value of the parameter is.

5% tests are the standard default for the significance level, because they usually give the best balance between the two possible errors.

However, there are occasions where reducing the size of Type I or Type II errors is important.

You will be given a particular level of significance in any case where you are expected to use something other than 5%.

Using a test with significance level of 1% reduces the probability of a Type I error but will give a higher probability of making a Type II error.
Conversely, if you want to reduce the likelihood of making a Type II error, you need to increase the significance level used to more than 5%.

Some diagrams will help to illuminate what happens with type II errors – remember that a test produces a critical region in which the null hypothesis is rejected, and all other outcomes are in the acceptance region. If the null hypothesis is not true, then p in the Binomial or λ in the Poisson take a value other than that specified in the null hypothesis, and P (Type II error) will be the probability of an observation falling in the acceptance region. Consider the two tailed test in the last section with $\lambda = 7.5$ - the blue bars represent outcomes in the critical (rejection) region and the maroon bars represent outcomes in the acceptance region.

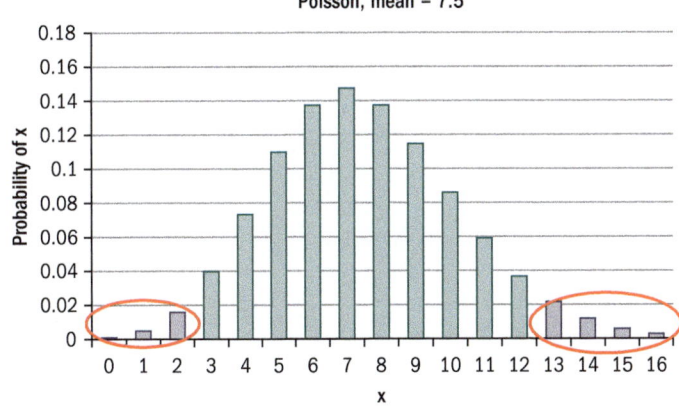

The highlighted outcomes are the rejection region under the null hypothesis.

And consider what the distribution looks like with $\lambda = 5$, keeping the colours for acceptance and critical regions the same:

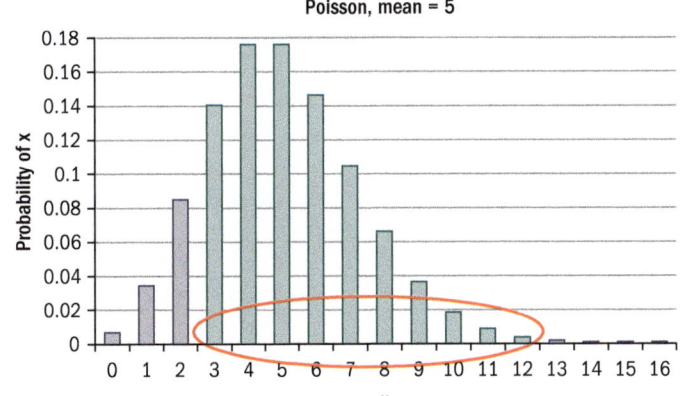

Here the null hypothesis is not true and a Type II error occurs if the observed outcome lies in the acceptance region for the test – the acceptance region is highlighted here.

Even though 5 feels a long way off the null value of 7.5, you can see how much of the distribution still falls within the acceptance region – leading to an incorrect conclusion. P(Type II error) = 0.733

In the last section you saw that a one tailed test for the Binomial with $n = 15$, with $H_0: p = 0.5$ vs $H_1: p < 0.5$ has critical region 0 – 3, giving the size of the test as 1.8% (which is the probability of a type I error). The distribution under H_0 showing the critical region in blue is shown below.

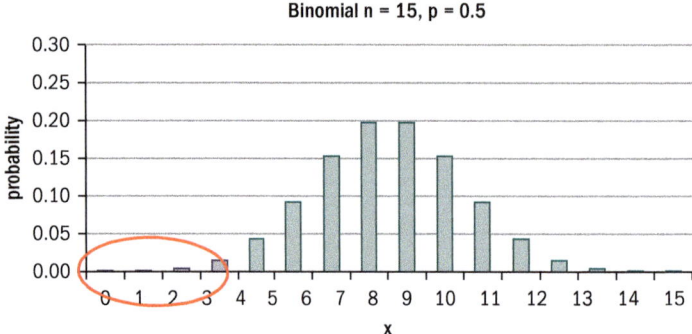

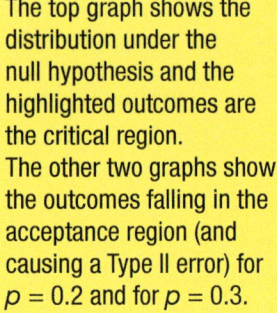

The top graph shows the distribution under the null hypothesis and the highlighted outcomes are the critical region.
The other two graphs show the outcomes falling in the acceptance region (and causing a Type II error) for $p = 0.2$ and for $p = 0.3$.

If you now consider a specific case of $p = 0.2$ (so only one success in five trials rather than one in two) you might feel that is a pretty big shift and that the test should identify it. In this case P(Type II error) is 35.1%

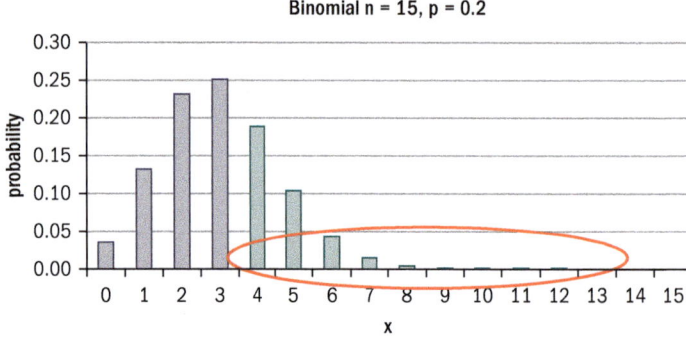

Moreover, as you move closer to the null hypothesis value, the probability of a type II error gets larger very quickly.
With $p = 0.3$, P(Type II error) is over 70%.

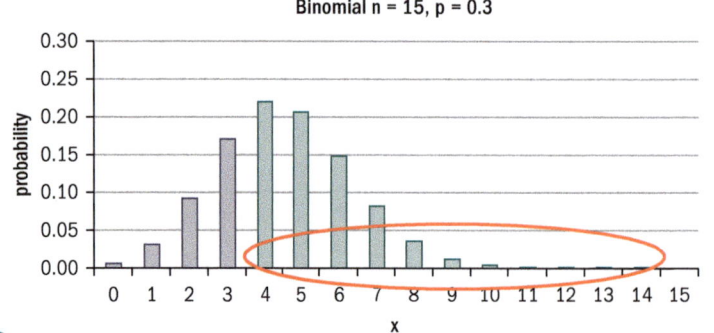

Computer software can calculate these Type II error probabilities for large numbers of possible alternative parameter values – giving a function known as the **power of the test** – which is really how good it is at discriminating the null hypothesis from the range of alternatives.

You don't have to deal with the power of the test – but it is why (intuitively) Vanya is going to be able to identify a biased coin with 500 tosses much better than Anil could with only five tosses.

Example 3

A test is being carried out on a Binomial distribution with $n = 15$, for H_0: $p = 0.3$, with a critical region of 0 and 1.

i) calculate the significance level of the test

ii) state the probability that a Type I error occurs

iii) if in fact $p = 0.1$, calculate the probability of a Type II error.

i) If $X \sim B(15, 0.3)$ with critical region 0 and 1 then the significance level is
$P(X = 0,1) = 0.7^{15} + 15 \times 0.7^{14} \times 0.3 = 0.00475 + 0.03052 = 0.035$.

ii) The probability that a Type I error occurs = the significance level (by definition) so it is 3.5%.

iii) If $X \sim B(15, 0.1)$ then the probability of a Type II error is the probability X is observed in the acceptance region i.e. $P(X > 1)$
$P(X > 1) = 1 - (0.9^{15} + 15 \times 0.9^{14} \times 0.1) = 1 - (0.20589 + 0.34315) = 0.451$
or 45.1%

Example 4

A test is being carried out on a Poisson distribution H_0: $\lambda = 4.5$ against H_1: $\lambda < 4.5$ with critical region of 0.

a) Calculate the significance level of the test.

b) When the test is carried out, the value observed is 3.

  i) State the conclusion of the test

  ii) State which type of error in the hypothesis test (Type I or Type II) could have been made in these circumstances.

a) If $X \sim P(4.5)$ then $P(X = 0) = e^{-4.5} = 0.011109 = 1.1\%$ so the significance level is 1.1% (the probability of being in the critical (rejection) region when H_0 is true).

b) i) Since 3 does not lie in the critical region, the decision is to accept H_0 and conclude that there is insufficient evidence to suggest that the mean is less than 4.5.

  ii) Since the decision is to accept H_0 the only error possible is the Type II where you incorrectly accept the null hypothesis.

Exercise 8.3

1. A test is being carried out on a Binomial distribution with $n = 15$, for $H_0: p = 0.2$, with a critical region of 0 and 1.

    i) Calculate the significance level of the test.

    ii) State the probability that a Type I error occurs.

    iii) If in fact $p = 0.1$, calculate the probability of a Type II error.

2. A test is being carried out on a Poisson distribution $H_0: \lambda = 6$ against $H_1: \lambda < 6$. A 2% test is to be carried out.

    i) Find the critical region for the test.

    ii) Calculate the actual significance level of the test.

    iii) State the probability that a Type I error occurs.

    iv) If in fact $\lambda = 3$, calculate the probability of a Type II error.

3. A test is being carried out on a Binomial distribution with $n = 20$, for $H_0: p = 0.3$, with a critical region of 0, 1 and 2.

    i) Calculate the significance level of the test.

    ii) State the probability that a Type I error occurs.

    iii) If in fact $p = 0.1$, calculate the probability of a Type II error.

4. A test is being carried out on a Poisson distribution $H_0: \lambda = 0.8$ against $H_1: \lambda > 0.8$. A 5% test is to be carried out.

    a) i) Find the critical region for the test.

    ii) Calculate the probability that a Type I error occurs.

    b) When the test is carried out, the value observed is 3.

    i) State the conclusion of the test.

    ii) State which type of error in the hypothesis test (Type I or Type II) could have been made in these circumstances.

5. A test is being carried out on a Poisson distribution $H_0: \lambda = 3$ against $H_1: \lambda < 3$.

    i) Show that the critical region for a 5% test contains only 0.

    ii) Show that the probability of a type II error for $\lambda = 0.3$ is greater than 25%.

6. A test is being carried out on a Binomial distribution with $n = 20$, for $H_0: p = 0.6$, against $H_1: p > 0.6$.

    a) i) Show that the critical region for a 5% test is $X \geq 17$.

    ii) State the probability that a Type I error occurs.

    b) If the proportion observed in the test was 75%, state which type of error in the hypothesis test (Type I or Type II) could have been made.

8.4 Hypothesis test for the proportion p of a Binomial distribution

Sections 8.4 and 8.5 look at doing the full hypothesis test procedure for the Binomial and the Poisson distributions.

This means identifying formally the null and alternative hypotheses, constructing the test as in section 8.2, and then applying the decision rule and stating the conclusion. In some cases you will also be asked about errors.

The conclusions will be in one of these forms:

- Reject H_0, and conclude that there is significant evidence (at the ...% level) that p is greater than ....
  (or whatever the appropriate context is).
- Accept H_0, and conclude that there is not significant evidence that p is greater than...
  This is a lack of evidence against the hypothesis, not evidence in support of it, and your language must reflect this.

> Take care with the language you use.

Example 5

30% of customers in a large store present the store's loyalty card when they buy something in the store.

The store runs an advertising campaign to promote their loyalty card. After the advertising campaign has finished, a random sample of 20 sales is examined and a loyalty card was used in 10 of them.

Test at the 5% level of significance whether the campaign has been effective in increasing the use of the loyalty card.

···

Let X = the number of sales where a loyalty card is used (in the random sample).

Then $X \sim B(20, p)$.

This is a 1-tailed test since it is looking for p to be increased from 0.3

The formal statement of the hypotheses is

$H_0: p = 0.3$ $\qquad$ $H_1: p > 0.3$.

> You are looking at the probability of 'as extreme (or more) than the critical value' so here.
> $P(X \geq 10) = 1 - P(X < 10)$
> $\qquad = 0.048$

Critical region method:

The closest you can get to a 5% test is a 4.8% test where the critical region is $X \geq 10$.

There were actually 10 cards used, so the decision is to reject H_0.

There is significant evidence (at the 5% level) to suggest that the advertising campaign has increased the use of the store loyalty card.

> You should always state the decision clearly.
> You should also state the conclusion in words in the context of the problem.

Probability method:

$P(X \geq 10) = 4.8\%$. This is <5%, so reject H_0 (and state the conclusion as before)

The value of 4.8% is known as the p – value of the observations – the probability of observing an outcome as least as extreme as what has been seen, if H_0 was true.

Most academic research results are reported quoting p – values, but they are not included in this course.

Example 6

A psychologist moves to New Zealand after working in the UK.

In the UK 25% of her patients suffered from a fear of heights.

In a random sample of 20 of her patients in New Zealand she finds that there are 7 who suffer from a fear of heights.

Test at the 5% level of significance whether there is a difference in proportions of people who suffer from a fear of heights in the UK and in New Zealand.

The psychologist has no reason to suppose that the proportion should be higher or lower in New Zealand than in the UK.

This should be a 2-tailed test looking for a difference.

The formal statement of the hypotheses is

$H_0: p = 0.25$ $H_1: p \neq 0.25$

Probability method:

$P(X > 7) = 10.2\% > 2.5\%$, so accept H_0 and conclude that there is not sufficient evidence at the 5% level to suggest the proportion in New Zealand is different from in the UK.

Do not think that because 7 out of 20 is >25% it is a 1-tailed test for an increase. The reason for a one tail test must not solely come from the data.

The comparison is with 2.5% because it is a 2 tailed 5% test.

Critical region method:

$P(X \leq 1) = 2.4\%$; $P(X \geq 10) = 14\%$ so the test is reject H_0 if X is 0, 1 or at least 10.

The observed outcome was 7 which does not lie in the critical region, so accept H_0 (and state the same conclusion).

Note that you can be asked explicitly to construct a critical region, so you must be able to conduct tests using that method. In cases like example 6 where you are conducting a two tailed test, there is less work involved in using the probability method if you are only asked to carry out the test for a discrete distribution – because you only need to investigate one tail of the distribution – but you must be careful to show that you are comparing the probability with half the size of the test.

Exercise 8.4

1. In a multiple choice examination paper in Psychology, a candidate has to select which one of five possible answers to a question is correct. On a paper of 15 questions she gets 6 correct answers.

   Her teacher says she can't have done any work at all. Test, at the 5% significance level, whether there is evidence to suggest that she has done better than just guessing.

2. A coin is tossed 12 times, and 10 heads are seen.

   i) Test at the 5% significance level whether the coin is biased.

   ii) State which type of error in the hypothesis test (Type I or Type II) could have been made in these circumstances.

3. An airline claims that at least 96% of their flights arrive within 5 minutes of their published arrival time. In a random sample of 30 flights three arrive more than 5 minutes after the published arrival time. Test the airline's claim at the 10% level of significance.

4. A train company claims that at least 90% of their trains arrive on time. In a random sample of 35 train arrivals, three arrive late. Explain why you would not need to do any detailed calculations in order to conclude this sample would not provide evidence to suggest that the company overstates their performance.

5. A financial advisor's publicity claims that at least 90% of their customers are satisfied with their performance.

   i) Is a 5% or a 10% test more likely to conclude that the financial advisor has overstated their performance?

   In a random sample of 20 customers 4 said they were not satisfied.

   ii) Carry out a 5% test to investigate whether the financial advisor has overstated their performance.

   iii) could a type II error have been made in this situation?

6. In a hotel 25% of people take longer than 10 minutes to register on arrival. The management install a new computer system which they claim will reduce the time to register. It is decided to accept the claim if, in a random sample of 24 people, the number taking longer than 10 minutes to register is not more than 2.

   i) Calculate the significance level of the test.

   ii) State the probability that a Type I error occurs.

   iii) Calculate the probability that a Type II error occurs if the probability of a person taking longer than 10 minutes to register is now 10%.

8.5 Hypothesis test for the mean of a Poisson distribution

You use the same process for testing the Poisson distribution:
identify the hypotheses formally and define the critical region;
carry out the test; state conclusions OR use the probability approach.

Example 7

A single observation x is to be taken from a Poisson distribution with parameter λ.

This observation is to be used to test $H_0: \lambda = 6$ against $H_1: \lambda \neq 6$.

i) Using a 5% significance level, find the critical region for this test assuming that the probability of rejection in either tail is to be no more than 2.5%.

ii) Write down the actual significance level of this test.
The actual value of x obtained was 2.

iii) State a conclusion that can be drawn based on this value.

....

i) $P(X \leq 1) = 1.7\%$ at the bottom end and $P(X \geq 12) = 2.0\%$ at the top end so the decision rule is: Reject H_0 if $X \leq 1$ or $X \geq 12$.

ii) The size of the test is $1.7\% + 2.0\% = 3.7\%$

iii) Since 2 is not in the critical region, you should accept H_0 and conclude that there is not sufficient evidence at the 5% level to suggest that the mean is not 6.

> In this case you are explicitly required to find the critical region so you would not then go to the probability method to carry out the test.

Example 8

The number of accidents on a stretch of road may be modelled by a Poisson distribution with an average rate of 2.5 accidents per week.

a) Find the probability that there are at least 5 accidents in a randomly chosen week.

Some new warning notices are installed and in a 3-week period there are 4 accidents.

b) Using a 5% significance level, test whether there has been a reduction in the average rate of occurrence of accidents.

....

a) if $X \sim P(2.5)$, $P(X \geq 5) = 1 - P(X \leq 4) = 1 - 0.8912 = 0.1088$

b) Test $H_0: \lambda = 7.5$ against $H_1: \lambda < 7.5$
(for the 3-week period the mean is $3 \times 2.5 = 7.5$ under H_0).

Probability approach:
$P(X \leq 4) = 0.132 > 5\%$, so accept H_0 and conclude there is not sufficient evidence to suggest that there has been a reduction in the average rate of accidents occurring.

Critical region:
$P(X = 0) = 0.0005$; $P(X \leq 1) = 0.0047$; $P(X \leq 2) = 0.0203$;
$P(X \leq 3) = 0.059 > 5\%$
So the critical region contains 0, 1 and 2. Since 4 does not lie in this region H_0 would be accepted as before.

Example 9

Luigi was recently appointed to be responsible for the service in a restaurant. During the previous year, the restaurant received an average of 3 emails per week complaining about the quality of service in the restaurant.

The number of such emails may be modelled by a Poisson distribution.

a) During the week before Luigi's appointment, 6 such emails were received. Examine, at the 5% level of significance, whether there is significant evidence that, immediately before Luigi's appointment, the mean number of such emails received exceeded 3 per week.

b) On his appointment, Luigi introduced changes to the methods of waiters recording orders and passing them to the kitchen. Following these changes, 2 emails of complaint were received during a two-week period. Examine, using a 5% level of significance, whether there is significant evidence that, following the changes introduced by Luigi, the mean number of such emails received was less than 3 per week.

c) Comment on the effectiveness of the changes introduced by Luigi.

a) If X is the number of emails received in a week then $X \sim P(3)$,
$P(X \geq 6) = 1 - P(X \leq 5) = 1 - 0.916 = 0.084 > 5\%$, so accept H_0 and conclude there is not sufficient evidence to suggest that the mean is more than 3 per week.

b) If Y is the number of emails received in two weeks then $Y \sim P(6)$,
$P(Y \leq 2) = 0.062 > 5\%$, so accept H_0 and conclude there is not sufficient evidence to suggest that the mean is less than 3 per week.

c) The number of complaints has gone from 6 in a week to only 2 in a two-week period, which looks as though the changes might have had an effect. However, the tests on the mean being higher than 3 per week before and being lower than 3 per week after did not provide sufficient evidence in either case, highlighting that accepting the null hypothesis in a test is a fairly weak statement – the problem is that testing in only one or two weeks makes it very difficult to identify a shift in the mean rate. If Luigi monitored the complaints over a period of two months he would have a much better evidence base on which to judge the effects of the changes.

Exercise 8.5

1. A single observation x is to be taken from a Poisson distribution with parameter λ. This observation is to be used to test H_0: $\lambda = 9.5$ against H_1: $\lambda < 9.5$.

   i) Using a 5% significance level, find the critical region for this test.

   ii) Write down the significance level of this test.

   The actual value of x obtained was 4.

   iii) State a conclusion that can be drawn based on this value.

2. "Hits" on a webpage at a particular location on the Internet occur at an average rate of one every three minutes, between the times 1100 and 1600.

   i) If X is the number of hits observed in a half hour period between the times 1100 and 1600, which distribution is appropriate to model this situation and why?

   ii) Calculate the probability that there are no more than 4 hits in this half hour.

   iii) Why might the probability that there are no more than 4 hits in the half hour 0300 to 0330 be different from the answer in b.

   iv) If there were 4 hits between 0300 and 0330, test at the 5% level of significance that there is a difference in the average rate of occurrences between 1100 to 1600 and 0300 to 0330.

3. Accidents occur at a certain road junction at the average of 3 per month.

   a) Suggest a suitable model for the number of accidents at the junction in the next month.

   b) Find the probability of 5 or more accidents at this road junction in the next month.

   The local residents have applied for a crossing to be installed. The planning committee agree to monitor the situation for the next 12 months. If there is at least one month with 5 or more accidents in it they will install a crossing.

   c) Find the probability that a crossing is installed.

   d) If a crossing is installed, and the local residents want to see if there is any improvement in the situation, give the decision rules which would apply if the test was to look at the number of accidents in

   i) a 1 month period;   **ii)** a 3 month period.

   e) Give 1 advantage of choosing the

   i) 1 month test period;   **ii)** 3 month test period.

4. A single observation x is to be taken from a Poisson distribution with parameter λ. This observation is to be used to test $H_0 : \lambda = 2.5$ against $H_1 : \lambda > 2.5$.

   i) Using a 5% significance level, find the critical region for this test

   ii) Write down the significance level of this test.

   The actual value of x obtained was 5.

   iii) State a conclusion that can be drawn based on this value.

5. Customers arrive at a store at an average rate of one every four minutes, between the times 0900 and 1600.

   i) if X is the number of customers in a ten minute period between the times 0900 and 1600, which distribution is appropriate to model this situation and why?

   ii) Calculate the probability that no more than 4 customers in this ten minute period.

   iii) Why might the probability that there are no more than 4 customers in the ten minutes just before the store closes be different from the answer in ii.

   iv) If 5 customers arrive in the ten minutes just before the store closes, test at the 5% level of significance that there is a difference in the average rate of occurrences between 0900 to 1600 and just before it closes.

6. Accidents occur on a certain section of a train route at the average of 2.2 per month.

a) Suggest a suitable model for the number of accidents on this section in the next month.

b) Find the probability of 5 or more accidents on this section in the next month.

The local residents have applied for a speed limit to be imposed on this route. The company agree to monitor the situation for the next 12 months. If there is at least one month with 5 or more accidents in it they will impose a speed limit on the section.

c) Find the probability that a speed limit is imposed.

d) If a speed limit is imposed, and the local residents want to see if there is any improvement in the situation,

    i) explain why it is not possible to have a test at the 5% level based on one month

    ii) give the decision which would apply if the (5%) test was to look a the total number of accidents in a 3 month period.

e) Give one other advantage of choosing the 3 month test period.

Summary exercise 8

1. For the following hypothesis tests, state whether you would use a 1 or 2 tailed test, and find the critical region and the exact significance level it will give

    i) 4. For a Binomial with $n = 10$, testing $H_0: p = 0.27$ against $H_1: p < 0.27$ with a 5% level of significance.

    ii) For a Poisson, testing $H_0: \lambda = 6.5$ against $H_1: \lambda \neq 6.5$ with a 10% level of significance.

2. A test is being carried out on a Poisson distribution $H_0: \lambda = 4.5$ against $H_1: \lambda < 4.5$. A 2% test is to be carried out.

    i) find the critical region for the test

    ii) calculate the actual significance level of the test

    iii) state the probability that a Type I error occurs

    iv) if in fact $\lambda = 2$, calculate the probability of a Type II error.

3. A city council claims that at least 95% of their residents are satisfied with the services the council provide. In a random sample of 20 residents, 4 said they were not satisfied.

    i) Carry out a 5% test to investigate whether the council has overstated the level of satisfaction with the services provided.

    ii) State what type of error could have been made in this test.

4. When a council published plans to redevelop an area in the centre of the town only 10% of local residents approved the plans. The council then modified the plans and, in a random sample of 20 local residents, 5 approved the modified plans.

    i) Show that a test at the 5% level of significance will accept that the proportion of local residents who approve of the modified plans is greater than for the original plans.

    ii) Do you think it would be a good idea for the council to go ahead with the modified plans on the basis of this increased approval rate?

5. It is claimed that a certain 6-sided die is biased so that it shows a one less often than if it was fair. In order to test this claim at the 5% level of significance the die is thrown 20 times and the number of ones is noted.

    i) Given that the die shows a one on 1 of the 20 throws, carry out the test.

    On another occasion the same test is carried out again.

    ii) Find the probability of a Type I error.

    iii) Explain what is meant by a Type II error in this context.

6. The number of hurricanes hitting a certain island over the past 150 years has followed a Poisson distribution with mean 1.2 per year. Insurance companies suspect that global warming has now increased the mean and they should increase their storm insurance premiums. A hypothesis test, at the 5% level of significance, is to be carried out to test this suspicion. The number of hurricanes which hit the island in the next year will be used for the test.

    i) Find the rejection region for the test.

    ii) Find the probability of making a Type II error if the mean number of hurricanes hitting the island has actually increased to 2.5 per year.

7. A hospital patient's white blood cell count has a Poisson distribution. Before undergoing treatment the patient had a mean white blood cell count of 5.2. After the treatment a random measurement of the patient's white blood cell count is made, and is used to test at the 10% significance level whether the mean white blood cell count has decreased.

    i) State what is meant by a Type I error in the context of the question, and find the probability that the test results in a Type I error. [4]

    ii) Given that the measured value of the white blood cell count after the treatment is 2, carry out the test. [3]

    iii) Find the probability of a Type II error if the mean white blood cell count after the treatment is actually 4.1. [3]

    Cambridge International A Level and AS Mathematics 9709 Paper 71 Q7 May / June 2010

8. In a certain city it is necessary to pass a driving test in order to be allowed to drive a car. The probability of passing the driving test at the first attempt is 0.36 on average. A particular driving instructor claims that the probability of his pupils passing at the first attempt is higher than 0.36. A random sample of 8 of his pupils showed that 7 passed at the first attempt.

    i) Carry out an appropriate hypothesis test to test the driving instructor's claim, using a significance level of 5%. [5]

    ii) In fact, most of this random sample happened to be careful and sensible drivers. State which type of error in the hypothesis test (Type I or Type II) could have been made in these circumstances and find the probability of this type of error when a sample of size 8 is used for the test. [4]

    Cambridge International A Level and AS Mathematics 9709 Paper 7 Q4 May / June 2009

9. Every month Susan enters a particular lottery. The lottery company states that the probability, p, of winning a prize is 0.0017 each month. Susan thinks that the probability of winning is higher than this, and carries out a test based on her 12 lottery results in a one-year period. She accepts the null hypothesis $p = 0.0017$ if she has no wins in the year and accepts the alternative hypothesis $p > 0.0017$ is she wins a prize in at least one of the 12 months.

    i) Find the probability of the test resulting in a Type I error. [2]

ii) If in fact the probability of winning a prize each month is 0.0024, find the probability of the test resulting in a Type II error. [3]

iii) Use a suitable approximation, with $p = 0.0024$, to find the probability that in a period of 10 years Susan wins a prize exactly twice. [3]

Cambridge International A Level and AS Mathematics 9709 Paper 7 Q5 November 2008

10. The number of injuries per month at a certain factory has a Poisson distribution. In the past the mean was 2.1 injuries per month. New safety procedures are put in place and the management wishes to use the next 3 months to test, at the 2% significance level, whether there are now fewer injuries than before, on average.

i) Find the critical region for the test. [5]

ii) Find the probability of a Type I error. [1]

iii) During the next 3 months there are a total of 3 injuries. Carry out the test. [3]

iv) Assuming that the mean remains 2.1, calculate an estimate of the probability that there will be fewer than 20 injuries during the next 12 months. [5]

Cambridge International A Level and AS Mathematics 9709 Paper 71 Q6 June 2011

Chapter summary

- The null hypothesis will be in the form $H_0 : p =$ or $H_0 : \lambda =$ and the alternative hypothesis is in one of the forms $H_1: p >$ or $p <$ or $p \neq$ (or similar for the Poisson).
- If the alternative hypothesis is λ or $p \neq$ then the test is two-tailed. The other forms give a one-tailed test.
- The significance level of the test is the probability that a test will reject the null hypothesis when it is actually true.
- Type I errors occur when the null hypothesis is incorrectly rejected – so P(Type I error) = significance level in any test.
- Type II errors occur when the null hypothesis is incorrectly accepted. The probability of a Type II error occurring depends on the particular value of the parameter but it is the probability that the distribution with that parameter value produces an observation lying in the acceptance region.
- Conclusions for a hypothesis need to refer to the evidence – there is evidence to suggest the null hypothesis is false, or a lack of evidence to suggest it is false (when the null is to be accepted) and it is important that conclusions do not imply something has been proved, nor that there is positive evidence in support of the null hypothesis.
- The critical region (or rejection region) for a hypothesis test are the set of values of the test statistic for which the null hypothesis will be rejected. The critical value(s) are the boundary value(s) of the critical region.
- The acceptance region for a hypothesis test are the set of values of the test statistic for which the null hypothesis will be accepted.
- Hypothesis tests for the Binomial and Poisson distributions are carried by comparing the observed value with the critical region, or by calculating the likelihood of seeing an observation at least as extreme at what was obtained, and comparing that probability with the significance level of the test.

9 Hypothesis testing using the Normal distribution

The rise of technology now means that there are often such large amounts of data that hypothesis testing is not really helpful – because with very large n almost every difference is significant even if it is not large enough to be an important difference, but there are still lots of applications like control systems for productions lines which are firmly rooted in the hypothesis testing framework.

Objectives

After studying this chapter you should be able to:

- formulate hypotheses and carry out a hypothesis test concerning the population mean in cases where the population is normally distributed with known variance or where a large sample is used;

- calculate the probabilities of making Type I and Type II errors in specific situations involving tests based on a normal distribution.

Before you start

You should know how to:

1. Calculate probabilities for the Normal distribution.
   If $X \sim N(83.2, 10.2)$ find the value of x for which $P(X < x) = 0.025$

   $$P(X < x) = P\left(Z < \frac{x - 83.2}{\sqrt{10.2}} \right) = 0.025$$

   $$\Rightarrow \frac{x - 83.2}{\sqrt{10.2}} = -1.96 \Rightarrow x = 76.9$$

Skills check:

1. If $X \sim N(57,12)$ find the value of x for which $P(X > x) = 0.025$

9.1 Hypothesis test for the mean of a Normal distribution

In chapter 8 you met the logical basis for hypothesis testing, where the sampling distribution of a statistic was used to identify what would be the most unlikely events to be observed **if the null hypothesis was true**. This set of outcomes then formed the critical region for the test. In chapter 4 and chapter 7 you developed the distribution of the sample mean when the population was a Normal distribution.

Putting these together we can set out how to test for the mean of a Normal distribution. As before, the construction of the test has a number of important steps in it:

- Identify formally the null and alternative hypotheses.
  - o The null hypothesis must give a distribution i.e. in this context it is always in the form $H_0: \mu =$
  - o The alternative hypothesis is in one of the forms: $H_1: \mu >$ or $\mu <$ or $\mu \neq$ (the first two are the one-tailed tests and the third is the two-tailed test)
- Give an explicit statement of the decision rule to be used.
  - o There are two standard formats for this:
    - ➤ State the critical value, and then work out the z-score for your observed value i.e. calculate the value of $z = \dfrac{\bar{x} - \mu}{\dfrac{\sigma}{\sqrt{n}}}$. So for a 5% level of significance the critical values will be as follows – two tailed test is $z = \pm 1.96$ and for the one tailed tests it is $z = 1.645$ or $z = -1.645$.
    - ➤ Or: "Accept H_0, at the 5% level of significance if $\bar{x} < \mu + 1.645 \times \dfrac{\sigma}{\sqrt{n}}$" if the alternative hypothesis is for $\mu >$. If the alternative hypothesis is $\mu \neq$ then the test will be "Accept H_0, at the 5% level of significance if $\bar{x} \in \left(\mu - 1.96 \times \dfrac{\sigma}{\sqrt{n}}, \mu + 1.96 \times \dfrac{\sigma}{\sqrt{n}} \right)$"
- Apply the decision rule to the value of $\bar{x}$ or z.
- Give the conclusions of the test in the context of the problem.
  - o The statement of the conclusions must avoid categorical statements. A test allows you to state either
  - 1 Accept H_0. This is **not because the evidence supports it**, and should not be reported as such. You can only conclude there is not significant evidence against it.
  - 2 Reject H_0 Conclude that there is evidence, significant at the 5% level (or whatever the level was), that the mean is not — (or of an increase/ decrease in mean if it is one-tailed).

Note: if you are told that the sample was discovered not to be random, then the conclusion of your test is invalidated.

It is worth looking at the distributions of a number of sampling distributions to highlight the importance of the size of the sample in being able to discriminate where there has been a shift in the mean. All of the diagrams below use the same scale (so the area under all the curves remains at one) so you can see what difference increasing the sample size has:

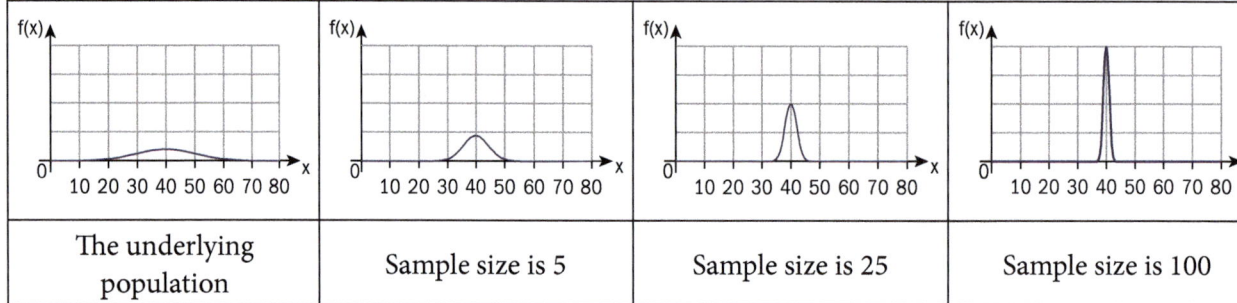

| The underlying population | Sample size is 5 | Sample size is 25 | Sample size is 100 |

One important aspect to realise is not only does the sampling distribution under H_0 shrink in width (so the acceptance region becomes much narrower) but for any particular value of the alternative hypothesis the sampling distribution has the same profile as these do, but centred on the other value – so the probability of the observed sample mean now falling in the acceptance region (which would be a Type II error) reduces very quickly indeed.

> ### Example 1
>
> A machine has produced bolts over a long period of time, where the length in millimetres was distributed as N(20.0, 0.16). It is believed that recently the mean length has changed. To test this a random sample of 10 bolts is taken and the mean length is found to be 19.8 mm. Carry out a hypothesis test at the 5% significance level to test whether the population mean has changed, assuming the variance remains the same.
>
> So the test is: $H_0: \mu = 20.0$ vs $H_1: \mu \neq 20.0$, with $\sigma = 0.4$ and $n = 10$
>
> **Method 1:** The critical values for the 5% two-tailed test are $z = \pm 1.96$
>
> $z = \dfrac{\bar{x} - \mu}{\frac{\sigma}{\sqrt{n}}} = \dfrac{19.8 - 22.0}{\frac{0.4}{\sqrt{10}}} = -1.581$, which lies in the acceptance region (between -1.96 and 1.96).
>
> **Method 2:** Accept H_0, at the 5% level of significance if $\bar{x} \in \left(\mu - 1.96 \times \dfrac{\sigma}{\sqrt{n}}, \mu + 1.96 \times \dfrac{\sigma}{\sqrt{n}} \right)$
>
> So accept H_0 if $\bar{x} \in \left(20.0 - 1.96 \times \dfrac{0.4}{\sqrt{10}}, 20.0 + 1.96 \times \dfrac{0.4}{\sqrt{10}} \right) = (19.752, 20.248)$
>
> Since 19.8 does lie in this interval you accept H_0.
>
> In either case the conclusion is 'Accept H_0: there is not sufficient evidence to suggest that the mean has changed.'

Example 2

The amount of rainfall on a certain island over the past 30 years has followed a Normal distribution with mean 82.3 cm per year, and standard deviation 15.3. Scientists suspect that global warming has now increased the mean. A hypothesis test, at the 5% level of significance, is to be carried out to test this suspicion. The average rainfall on the island over the next 5 years will be used for the test.

a) i)   Find the rejection region for the test.

   ii) What is the probability of making a Type I error?

   iii) Find the probability of making a Type II error if the mean rainfall on the island has actually increased to 105 cm per year.

b)   One of the scientists believes 5 years is too long to wait to carry out this test, and proposes to carry out the test using the rainfall in one year.

   i)   Find the rejection region for this test.

   ii) Find the probability of making a Type II error now.

a)   So the test is: $H_0 : \mu = 82.3$ vs $H_1 : \mu > 82.3$, with $\sigma = 15.3$ and $n = 5$

   i)   The critical values for the 5% one tailed test are $z = 1.645$ so the rejection region is
$$\bar{x} > 82.3 + 1.645 \times \frac{15.3}{\sqrt{5}} = 93.6$$

   ii) because the Normal distribution is continuous, the probability of a Type I error for a 5% test is always 5%

   iii) if $X \sim N(105, 15.3^2)$, then
$$P(x < 93.6) = P\left(z < \frac{93.6 - 105}{\frac{15.3}{\sqrt{5}}} = -1.666\right) = 1 - \Phi(1.666) = 0.048$$

   so the probability of a Type II error would 4.8% for this test.

b) i)   The critical value for the 5% one tailed test is still $z = 1.645$ but the standard deviation of the test statistic is now 15.3, so the rejection region is $x > 82.3 + 1.645 \times 15.3 = 107.5$.

   ii) if $X \sim N(105, 15.3^2)$, then
$$P(x < 107.5) = P\left(z < \frac{107.5 - 105}{15.3} = 0.163\right) = \Phi(0.163) = 0.565$$

   so the probability of a Type II error would 56.5% for this

> The chance of type I error will be the same but using a single year dramatically reduces the capacity of the test to discriminate important changes in mean.

Exercise 9.1

1. For each of the following situations carry out the appropriate test using one of the methods listed in example 1 above.

   i) $X \sim N(\mu, 16)$. $H_0 : \mu = 38.3$ vs $H_1 : \mu \neq 38.3$.   5% test.
   $n = 20$.   DATA: $\bar{x}_{20} = 40.6$

   ii) $X \sim N(\mu, 40)$. $H_0 : \mu = 38.3$ vs $H_1 : \mu \neq 38.3$.   5% test.
   $n = 20$.   DATA: $\bar{x}_{20} = 40.6$

   iii) $X \sim N(\mu, 40)$. $H_0 : \mu = 38.3$ vs $H_1 : \mu \neq 38.3$.   5% test.
   $n = 60$.   DATA: $\bar{x}_{60} = 40.6$

2. For each of the following situations carry out the appropriate test using the other method listed in example 1 above.

   i) $X \sim N(\mu, 19)$. $H_0 : \mu = 81.3$ vs $H_1 : \mu < 81.3$.   5% test.
   $n = 20$.   DATA: $\bar{x}_{20} = 79.5$

   ii) $X \sim N(\mu, 34)$. $H_0 : \mu = 81.3$ vs $H_1 : \mu < 81.3$.   5% test.
   $n = 20$.   DATA: $\bar{x}_{20} = 79.5$

   iii) $X \sim N(\mu, 34)$. $H_0 : \mu = 81.3$ vs $H_1 : \mu < 81.3$.   5% test.
   $n = 60$.   DATA: $\bar{x}_{60} = 79.5$

3. Look at the differences between the three tests in each of questions 1 and 2.

   i) How does the change in variance affect the outcome of the test when everything else is the same?

   ii) How does the change in sample size affect the outcome of the test when everything else is the same?

4. Jars of jam are filled by a machine. It has been found that the quantity of jam in a jar is normally distributed and has mean 351.2 g, with standard deviation 4.1 g. It is believed that the settings of the mean amount on the machine might have been altered accidentally. A random sample of 40 jars is taken and the mean quantity per jar is found to be 349.9 g. Assuming that the standard deviation has not been altered, state suitable null and alternative hypotheses and carry out a test using a 5% level of significance.

5. The systolic blood pressure of healthy young adults can be modelled by a Normal distribution with mean 105 and standard deviation 7. It is thought that otherwise healthy young adults who do not eat the recommended amounts of fruit and vegetables may have higher blood pressure than usual.

A random sample of 10 young adults who eat low amounts of fruit and vegetables but are otherwise healthy is taken and their systolic blood pressures are recorded here:

119, 106, 104, 121, 116, 118, 108, 113, 108, 112

Test at the 2% level of significance whether young adults who eat low amounts of fruit and vegetables have higher levels of systolic blood pressure than usual for healthy young adults.

6. The usable life of batteries produced by a manufacturer is normally distributed with a mean of 85.3 hours and standard deviation 3.2 hours. The managing director wants to monitor production to ensure that the mean remains at 85.3 hours. His production manager suggests two monitoring plans:

Plan A: test random samples of 5 batteries

Plan B: test random sample of 30 batteries.

i) for each plan find the acceptance region for a 5% two-tailed test.

ii) for each plan, calculate the probability of a Type II error if the mean usable life shifts by 2 hours.

The managing director looks at this information and asks the production manager to make a plan for a 5% two tailed test for which the probability of a Type II error if the mean usable life shifts by 2 hours is not more than 2.5%.

Note: The last part of this question goes beyond the requirements of the syllabus.

iii) find the minimum sample size which will satisfy this condition.

9.2 Hypothesis test for the mean using a large sample

If however, you are told the population was not Normal, or even not told that it is so there is no reason why you should assume that it is, then the conclusion is still valid if the sample size is large enough to invoke the Central Limit Theorem [$n > 30$] – you should explicitly refer to the Central Limit Theorem, not just refer vaguely to 'large samples'. Also with large samples you may use the unbiased estimate of the variance (based on the data) if you don't know the population variance.

When the conditions are met for the Binomial or Poisson distributions to be approximated by the Normal distribution, you may set up a hypothesis test using the Normal distribution. In this case you do not need to use a continuity correction.

In the Poisson you use the sample mean to estimate the parameter λ, and use it for both the mean and variance of the approximating Normal distribution.

In the Binomial you use the sample proportion, p_s, as the estimate of the parameter p, and use $\dfrac{p_s q_s}{\sqrt{n}}$ as the standard error in constructing the critical values.

Example 3

A college thinks the average travel time for its students is about half an hour. A random sample of 80 students attending the college is taken and their travel times recorded in minutes. The data are summarised by $\sum t = 2624$; $\sum t^2 = 96283$

i) Calculate an unbiased estimate of the variance of the travel times.

ii) Test at the 5% level of the significance whether the mean travel time is half an hour.

i) $\bar{x} = \dfrac{2624}{80} = 32.8$; $s^2 = \dfrac{n}{n-1}\left\{\dfrac{\sum t^2}{n} - (\bar{t})^2\right\} = \dfrac{80}{79}\left(\dfrac{96283}{80} - 32.8^2\right) = 129.3$ is the unbiased estimate

of the variance.

ii) **Method 1**: The critical values for the 5% two tailed test are $z = \pm 1.96$

$z = \dfrac{\bar{x} - \mu}{\frac{s}{\sqrt{n}}} = \dfrac{32.8 - 30}{\frac{\sqrt{129.3}}{\sqrt{80}}} = 2.20 > 1.96$, so reject H_0 and conclude that there is evidence at the 5%

level of significance to suggest that the mean travel time is not half an hour.

Method 2: Accept H_0 at the 5% level of significance if

$\bar{t} \in \left(30 - 1.96 \times \dfrac{\sqrt{129.3}}{\sqrt{80}}, 30 + 1.96 \times \dfrac{\sqrt{129.3}}{\sqrt{80}}\right) = (27.5, 32.5)$. Since 32.8 does not lie in this interval

you reject H_0 (as in method 1).

Example 4

Each multiple choice question in a test has 5 suggested answers, exactly one of which is correct. Novak feels he knows enough about the subject matter that he can do better than just guessing randomly because one or two suggested answers are usually obviously wrong. There are 50 questions on the test and Novak gets 18 correct.

i) State null and alternative hypotheses for a test of Novak's claim.

ii) Using a Normal approximation, test at the 1% level of significance whether Novak's claim is justified.

i) if X is the number of questions Novak answers correctly then $X \sim B(50, p)$ and the hypotheses are $H_0 : p = 0.2$ *vs* $H_1 : p > 0.2$.

ii) Using the Normal approximation in a one-tailed 1% test (cv is 2.326) gives the test as 'Accept

H_0 if the proportion observed is $> 0.2 + 2.326 \times \sqrt{\dfrac{0.2 \times 0.8}{50}} = 0.332$

Novak gets 18 out of 50, or 36% correct, so reject H_0 and conclude that there is evidence at the 1% level of significance to suggest that Novak does better than he would if he was just guessing randomly.

Example 5

The number of a certain endangered species of frog found in a wetland area can be modelled by a Poisson distribution with an average rate of 0.013 frogs per square metre. Conservationists are concerned that the species is in decline in the area because of recent construction work in neighbouring areas.

They mark off a square in the wetland with sides of 100 metres, and count the number of frogs in the area to test if the numbers are in decline.

i) State null and alternative hypotheses for the test.

ii) Using a suitable approximation, find the critical region for a 5% test.

iii) If the mean rate has dropped to 0.009 frogs per square metre calculate the probability that a Type II error will occur.

....

i) for the area selected, 10 000 sq metres, the expected number of frogs is 130. So the hypotheses are $H_0 : \lambda = 130$ *vs* $H_1 : \lambda < 130$

ii) Using the N(130, 130) to approximate the Po(130) distribution, and using a 5% one tailed test the critical region is 'Reject H_0 if $x > 130 - 1.645 \times \sqrt{130} = 111.2$'.

iii) A Type II error occurs if the null hypothesis is accepted when it is not true. If the mean rate drops to 0.009, then the null hypothesis is not true and a Type II error can occur. In this case the distribution of the number of frogs observed in a square of side 100 metres would be Po(90), which be approximated by the N(90, 90) distribution. Then P(Type II error occurs)

$$= \text{Pr(at least 112 frogs are observed)} = P\left(z > \frac{111.5 - 90}{\sqrt{90}} = 2.266 \right) = 1 - 0.9883 = 0.0116$$

Exercise 9.2

1. The summary statistics of a number of random samples are given below. For each one, find the unbiased estimator of the variance and carry out the test specified.

    i) $H_0 : \mu = 45$ *vs* $H_1 : \mu \neq 45$.   5% test. $n = 40$.   DATA: $\Sigma x = 1846$; $\Sigma x^2 = 86835$

    ii) $H_0 : \mu = 57.5$ *vs* $H_1 : \mu < 57.5$.   5% test. $n = 63$.   DATA: $\Sigma x = 3514$; $\Sigma x^2 = 199281$

    iii) $H_0 : \mu = -0.5$ *vs* $H_1 : \mu > -0.5$.   5% test. $n = 51$.   DATA: $\Sigma x = 24.3$; $\Sigma x^2 = 994.27$

2. Using suitable approximations, carry out the appropriate tests for each of the situations described below:

    i) $X \sim B(145, p)$. $H_0 : p = 0.4$ *vs* $H_1 : p \neq 0.4$.   5% test.   DATA: $x = 52$

    ii) $X \sim Po(\lambda)$. $H_0 : \lambda = 27.1$ *vs* $H_1 : \mu < 27.1$.   1% test.   DATA: $x = 19$

    iii) $X \sim B(92, p)$. $H_0 : p = 0.8$ *vs* $H_1 : p > 0.8$.   2% test.   DATA: $x = 65$

3. The spokesperson for a group of insurance companies claims that less one in five homeowners has had an increase of more than 10% in the premium for home insurance.

A survey of 85 randomly selected homeowners asks if their home insurance premiums have increased by more than 10% at the last renewal and 22 say that it did. Carry out a test at the 5% level of significance of the claim by the spokesperson.

4. The number of accidents recorded at a dangerous corner has followed a Poisson distribution with an average rate of 1.2 per week. A large warning sign is erected shortly before the corner. Over the next 6 months (26 weeks) a total of 20 accidents are recorded at the corner. Carry out a 1% test to investigate whether the erection of the sign appears to have reduced the number of accidents.

5. A certain variety of bush grows to heights which are normally distributed with mean 74.0 cm. A new fertiliser is introduced in the hope that this will increase the heights. The nursery owner records the heights of a large random sample of n bushes, and calculates that $\bar{x} = 75.2$ and $s = 5.3$.

    i) She consults a friend who is a statistician as to whether or not there is evidence that the heights have increased. The friend calculates the test statistic, z, has a value of 1.867 correct to 3 decimal places. Calculate the value of n.

    ii) Using this value of the test statistic, carry out the test at the 5% level of significance.

6. The number of letters received by a household on a weekday follows a Poisson distribution with mean 3.5.

    a) What is the probability that
        i) on a particular weekday the household receives two or more letters;
        ii) the total number of letters received on ten successive weekdays is thirty or more?

       (You should use a suitable approximation.)

    b) Explain briefly why a Poisson distribution is unlikely to provide an adequate model for the number of letters received on a weekday throughout the year.

    c) On the Monday before Christmas, 7 letters were delivered to the household. Test at the 1% level of significance whether the mean number of letters per day is higher in the week before Christmas than normally.

9.3 Using a confidence interval to carry out a hypothesis test

Consider the process used in method 2 for a 2-tailed test, where the acceptance region is identified, and the test carried out by seeing if the observed sample mean lies within the interval centred on the (hypothesised) population mean, μ. If you construct the corresponding confidence interval (a 95% CI corresponds to a 5% significance level test) there is a 1-1 correspondence between occasions in which μ lies in the CI centred on $\bar{x}$ and when lies $\bar{x}$ in the acceptance region – which is centred on μ.

In situations where you obtain a confidence interval, you can then use it to test any claim made.

> ### Example 6
>
> A random sample of 40 bottles of Aguafresh gave unbiased estimates of 250.3 millilitres and 15.8 millilitres² for the population mean and variance.
>
> i) Calculate a symmetric 95% confidence interval for the population mean.
>
> ii) The manufacturer claims that the mean volume of Aguafresh bottles is 252 millilitres. State with a reason whether your answer to part i) supports this claim.

i)   the 95% CI is given by $\bar{x} \pm 1.96\dfrac{s}{\sqrt{n}}$

$= 250.3 \pm 1.96\dfrac{\sqrt{15.8}}{\sqrt{40}} = (249.1, 251.5)$

ii) 252 millilitres lies outside this CI so there is evidence
to suggest that the mean is not 252.

Exercise 9.3

1. For the following sets of summary statistics construct the confidence interval indicated.
Comment on the claim made in each case in the light of the confidence interval you
have calculated

a) $n = 50$; $\sum t = 257$; $\sum t^2 = 2433$ 90% CI   claim $\mu = 4.5$

b) $n = 117$; $\sum x = 8596$; $\sum x^2 = 2186591$ 95% CI   claim $\mu = 90$

c) $n = 82$; $\sum v = -459$; $\sum v^2 = 5109$ 99% CI   claim $\mu = -6$

2. A random sample of 150 bottles of fruit juice gave unbiased estimates of
600.8 millilitres and 26.8 millilitres² for the population mean and variance.

i)   Calculate a symmetric 95% confidence interval for the population mean.

ii) The manufacturer claims that the mean volume of fruit juice bottles is 602 millilitres.
State with a reason whether your answer to part i) supports this claim.

3. A random sample of 80 jars of pickle gave unbiased estimates of 248.5 grams
and 16.8 grams² for the population mean and variance.

i)   Calculate a symmetric 95% confidence interval for the population mean.

ii) The manufacturer claims that the mean contents of the jars is 250 grams.
State with a reason whether your answer to part i) supports this claim.

Summary exercise 9

1. Jars of mayonnaise are filled by a machine.
It has been found that the quantity of
mayonnaise in a jar is normally distributed
and has mean 452.7 g, with standard
deviation 6.1 g. It is believed that the settings
of the mean amount on the machine might
have been altered accidentally.

i)   The company might be concerned if the
mean has decreased because the labels
claim that the jars contain at least 450
grams. Give a reason why the company
might be concerned if the mean has
increased.

A random sample of 70 jars is taken and the
mean quantity per jar is found to be 447.9 g.

ii) Assuming that the standard deviation
has not been altered, state suitable null
and alternative hypotheses and carry out
a test using a 5% level of significance.

2. The pulse of healthy young adults can be
modelled by a Normal distribution with
mean 80 and standard deviation 9. It is
thought that trained athletes may have lower
pulse rates than usual.

A random sample of 12 trained athletes is taken and their pulses are recorded here: 48, 52, 59, 45, 57, 52, 44, 58, 56, 44, 63, 50 Test at the 5% level of significance whether trained athletes have lower pulse rates than normal healthy young adults.

3. i) Give a reason why, in carrying out a statistical investigation, a sample may be used rather than a complete population. The spokesperson for a group of travel companies claims that 40% of their holidays are cheaper this year than they were last year.
        A survey of 65 randomly selected customers asks if their holiday is cheaper this year than last year, and 15 of them said yes.
    ii) Show that a test at the 5% level of significance of the claim by the spokesperson using this data would reject the claim.
    iii) When challenged about the claim, using the test as evidence, the spokesperson says that people may be paying more for a better holiday this year and her claim was comparing like for like prices. Comment on whether this might be a valid defence of the claim made.

4. The number of accidents recorded on a campsite during past summers had followed a Poisson distribution with an average rate of 2.3 per week. The owner employed a lifeguard last summer in an effort to improve the safety record at the campsite. Over the 3 months the lifeguard works (13 weeks) a total of 18 accidents are recorded at the campsite. Carry out a 5% test to investigate whether employing the lifeguard appears to have reduced the number of accidents.

5. A certain species of rose grows to heights which are normally distributed with mean 64.0 cm. A hybrid is created which is hoped

will increase the heights. The breeder records the heights of a large random sample of 60 bushes, and calculates that $\bar{x} = 65.2$ and $s = 5.6$.

Test at the 1% level of significance whether the height of the hybrid is greater than that of the original species.

6. Each multiple choice question in a test has 4 suggested answers, exactly one of which is correct. Rehka knows nothing about the subject of the test, but claims that she has a special method for answering the questions that is better than just guessing. There are 60 questions in the test, and Rekha gets 22 correct.
    i) State null and alternative hypotheses for a test of Rehka's claim.   [1]
    ii) Using a normal approximation, test at the 5% significance level whether Rehka's claim is justified.   [4]

    Cambridge International A Level and AS Mathematics 9709 Paper 7 Q1 June 2004

7. A study of a large sample of books by a particular author shows that the number of words per sentence can be modelled by a normal distribution with mean 21.2 and standard deviation 7.3, A researcher claims to have discovered a previously unknown book by this author. The mean length of 90 sentences chosen at random in this book is found to be 19.4 words.
    i) Assuming the population standard deviation of sentence lengths in this book is also 7.3, test at the 5% level of significance whether the mean sentence length is the same as the author's. State your null and alternative hypotheses.   [5]
    ii) State in words relating to the context of the test what is meant by a Type I error and state the probability of a Type I error in the test in part **i**).   [2]

    Cambridge International A Level and AS Mathematics 9709 Paper 7 Q4 June 2005

8. The times taken for the pupils in Ming's year group to do their English homework have a normal distribution with standard deviation 15.7 minutes. A teacher estimates that the mean time is 42 minutes. The times taken by a random sample of 3 students from the year group were 27, 35 and 43 minutes. Carry out a hypothesis test at the 10% significance level to determine whether the teacher's estimate for the mean should be accepted, stating the null and alternative hypotheses. [5]

Cambridge International A Level and AS Mathematics 9709 Paper 71 Q2 November 2008

9. The time taken for Samuel to drive home from work is distributed with mean 46 minutes. Samuel discovers a different route and decides to test at the 5% level whether the mean time has changed. He tries this route on a large number of different days chosen randomly and calculates the mean time.
   i) State the null and alternative hypotheses for this test. [1]
   ii) Samuel calculates the value of his test statistic z to be -1.729. What conclusion can he draw? [2]

Cambridge International A Level and AS Mathematics 9709 Paper 7 Q1 November 2006

Chapter summary
- The logic of hypothesis tests with continuous distributions is the same as for discrete distributions but you will be able to construct a test with any specified level of significance.
  - The null hypothesis must give a distribution i.e. in this context it is always in the form $H_0: \mu =$
- The alternative hypothesis is in one of the forms: $H_1: \mu >$ or $\mu <$ or $\mu \neq$ (the first two are the one-tailed tests and the third is the two-tailed test)
- Give an explicit statement of the decision rule to be used.
  There are two standard formats for this:
  - State the critical value, and then work out the z-score for your observed value i.e. calculate the value of $z = \dfrac{\bar{x} - \mu}{\dfrac{\sigma}{\sqrt{n}}}$. So for a 5% level of significance the critical values will be as follows – two tailed test is $z = \pm 1.96$ and for the one tailed tests it is $z = 1.645$ or $z = -1.645$.
  - or: "Accept H_0, at the 5% level of significance if $\bar{x} < \mu + 1.645 \times \dfrac{\sigma}{\sqrt{n}}$" if the alternative hypothesis is for $\mu >$. If the alternative hypothesis is $\mu \neq$ then the test will be "Accept H_0, at the 5% level of significance if $\bar{x} \in \left(\mu - 1.96 \times \dfrac{\sigma}{\sqrt{n}}, \mu + 1.96 \times \dfrac{\sigma}{\sqrt{n}} \right)$"
- Apply the decision rule to the value of $\bar{x}$ or z.
- Give the conclusions of the test in the context of the problem.
- The mean of a large sample from a population not known to be Normal can be done using the Central Limit Theorem.
- For large samples from Binomial and Poisson distributions the Normal distribution may be used as an approximation if the required conditions are met, and the test carried out in the standard process.
- If an $\alpha\%$ confidence interval has been constructed for a population mean, μ, then a $(100 - \alpha)\%$ a level significance test can be carried out by inspection – by considering whether the hypothesised mean, μ, is captured in the $\alpha\%$ confidence interval.

Review exercise C

1. Packets of fish food have weights that are distributed with standard deviation 2.3 g. A random sample of 200 packets is taken. The mean weight of this sample is found to be 99.2 g. Calculate a 99% confidence interval for the population mean weight. [3]

   Cambridge International AS and A Level Mathematics 9709 Paper 7 Q1 June 2006

2. A consumer group, interested in the mean fat content of a particular type of sausage, takes a random sample of 20 sausages and sends them away to be analysed. The percentage of fat in each sausage is as follows.

26	27	28	28	28	29	29	30	30
31	32	32	32	33	33	34	34	34
35	35							

   Assume that the percentage of fat is normally distribution with mean μ, and that the standard deviation is known to be 3.
   i) Calculate a 98% confidence interval for the population mean percentage of fat. [4]
   ii) The manufacturer claims that the mean percentage of fat in sausages of this type is 30. Use your answer to part i) to determine whether the consumer group should accept this claim. [2]

   Cambridge International AS and A Level Mathematics 9709 Paper 7 Q3 June 2003

3. A random sample of n people were questioned about their internet use. 87 of them had a high-speed internet connection. A confidence interval for the population proportion having a high-speed internet connection is $0.1129 < p < 0.1771$.
   i) Write down the mid-point of this confidence interval and hence find the value of n. [3]
   ii) This interval is an $\alpha\%$ confidence interval. Find α. [4]

   Cambridge International AS and A Level Mathematics 9709 Paper 71 Q2 June 2010

4. The weights in grams of oranges grown in a certain area are normally distributed with mean μ and standard deviation σ. A random sample of 50 of these oranges was taken, and a 97% confidence interval for μ based on this sample was (222.1, 232.1).
   i) Calculate unbiased estimates of μ and σ^2. [4]
   ii) Estimate the sample size that would be required in order for a 97% confidence interval for μ to have width 8. [3]

   Cambridge International AS and A Level Mathematics 9709 Paper 71 Q2 June 2009

5. A manufacturer claims that 20% of sugar-coated chocolate beans are red. George suspects that this percentage is actually less than 20% and so he takes a random sample of 15 chocolate beans and performs a hypothesis test with the null hypothesis $p = 0.2$ against the alternative hypothesis $p < 0.2$. He decides to reject the null hypothesis in favouour of the alternative hypothesis if there are 0 or 1 red beans in the sample.
   i) With reference to this situation, explain what is meant by a Type I error. [1]
   ii) Find the probability of a Type I error in George's test. [3]

   Cambridge International AS and A Level Mathematics 9709 Paper 7 Q2 November 2005

6. In a research laboratory where plants are studied, the probability of a certain type of plant surviving was 0.35. The laboratory manager changed the growing conditions and wished to test whether the probability of a plant surviving has increased.
   i) The plants were grown in rows, and when the manager requested a random sample of 8 plants to be taken, the technician took all 8 plants from the front row. Explain what was wrong with the technician's sample. [1]

ii) A suitable sample of 8 plants was taken and 4 of these 8 plants survived. State whether the manager's test is one-tailed or two-tailed and also state the null and alternative hypotheses. Using a 5% significance level, find the critical region and carry out the test. [7]

iii) State the meaning of a Type II error in the context of the test in part **ii)**. [1]

iv) Find the probability of a Type II error for the test in part **ii)** if the probability of a plant surviving is now 0.4. [2]

Cambridge International AS and A Level Mathematics 9709 Paper 7 Q7 November 2004

7. A survey of a random sample of n people found that 61 of them read *The Reporter* newspaper. A symmetric confidence interval for the true population proportion, p, who read *The Reporter* is $0.1993 < p < 0.2887$.

i) Find the mid-point of this confidence interval and use this to find the value of n. [3]

ii) Find the confidence level of this confidence interval. [4]

Cambridge International A Level and AS Mathematics 9709 Paper 7 Q3 May / June 2005

8. i) Explain what is meant by the term 'random sample'. [1]

In a random sample of 350 food shops it was found that 130 of them had Special Offers.

ii) Calculate an approximate 95% confidence interval for the proportion of all food shops with Special Offers. [4]

iii) Estimate the size of a random sample required for an approximate 95% confidence interval for this proportion to have a width of 0.04. [3]

Cambridge International A Level and AS Mathematics 9709 Paper 71 Q3 November 2007

9. The volumes of juice in bottles of Apricola are normally distributed. In a random sample of 8 bottles, the volumes of juice, in millilitres, were found to be as follows.

332   334   330   328   331   332   329   333

i) Find unbiased estimates of the population mean and variance. [3]

A random sample of 50 bottles of Apricola gave unbiased estimates of 331 millilitres and 4.20 millilitres^2 for the population mean and variance respectively.

ii) Use this sample of size 50 to calculate a 98% confidence interval for the population mean. [3]

iii) The manufacturer claims that the mean volume of juice in all bottles is 333 millilitres. State, with a reason, whether your answer to part **ii)** supports this claim. [1]

Cambridge International A Level and AS Mathematics 9709 Paper 7 Q6 November 2006

10. The management of a factory thinks that the mean time required to complete a particular task is 22 minutes. The times, in minutes, taken by employees to complete this task have a normal distribution with mean μ and standard deviation 3.5. An employee claims that 22 minutes is not long enough for the task. In order to investigate this claim, the times for a random sample of 12 employees are used to test the null hypothesis $\mu = 22$ against the alternative hypothesis $\mu > 22$ at the 5% significance level.

i) Show that the null hypothesis is rejected in favour of the alternative hypothesis if $\bar{x} > 23.7$ (correct to 3 significant figures), where $\bar{x}$ is the sample mean. [3]

ii) Find the probability of a Type II error given that the actual mean time is 25.8 minutes. [4]

Cambridge International AS and A Level Mathematics 9709 Paper 71 Q5 November 2011

11. In order to obtain a random sample of people who live in her town, Jane chooses people at random from the telephone directory for her town.

   i) Give a reason why Jane's method will not give a random sample of people who live in the town. [1]

   Jane now uses a valid method to choose a random sample of 200 people from her town and finds that 38 live in apartments.

   ii) Calculate an approximate 99% confidence interval for the proportion of all people in Jane's town who live in apartments. [4]

   iii) Jane uses the same sample to give a confidence interval of width 0.1 for this proportion. This interval is an x% confidence interval. Find the value of x. [4]

Cambridge International AS and A Level Mathematics 9709 Paper 71 Q6 November 2012

12. A machine has produced nails over a long period of time, where the length in millimetres was distributed as N(22.0, 0.19). It is believed that recently the mean length has changed. To test this belief a random sample of 8 nails is taken and the mean length is found to be 21.7 mm. Carry out a hypothesis test at the 5% significance level to test whether the population mean has changed, assuming that the variance remains the same. [5]

Cambridge International AS and A Level Mathematics 9709 Paper 7 Q3 June 2007

13 At a certain airport 20% of people take longer than an hour to check in. A new computer system is installed, and it is claimed that this will reduce the time to check in. It is decided to accept the claim if, from a random sample of 22 people, the number taking longer than an hour to check in is either 0 or 1.

   i) Calculate the significance level of the test. [3]

   ii) State the probability that a Type I error occurs. [2]

   iii) Calculate the probability that a Type II error occurs if the probability that a person takes longer than an hour to check in is now 0.09. [3]

Cambridge International AS and A Level Mathematics 9709 Paper 7 Q4 June 2007

14. The number of cars caught speeding on a certain length of motorway is 7.2 per day, on average. Speed cameras are introduced and the results shown in the following table are those from a random selection of 40 days after this.

Number of cars caught speeding	4	5	6	7	8	9	10
Number of days	5	7	8	10	5	2	3

   i) Calculate unbiased estimates of the population mean and variance of the number of cars per day caught speeding after the speed cameras were introduced. [3]

   ii) Taking the null hypothesis H_0 to be $\mu = 7.2$, test at the 5% level whether there is evidence that the introduction of speed cameras has resulted in a reduction in the number of cars caught speeding. [5]

   iii) State what is meant by a Type I error in words relating to the context of the test in part ii). Without further calculation, illustrate on a suitable diagram the region representing the probability of this Type I error. [3]

Cambridge International AS and A Level Mathematics 9709 Paper 7 Q7 June 2006

15. Over a long period of time it is found that the time spent at cash with drawal points follows a normal distribution with mean 2.1 minutes and standard deviation 0.9 minutes. A new system is tried out, to speed up the procedure. The null hypothesis is that the mean time spent is the same under the new system as previously. It is decided to reject the null

hypothesis and accept that the new system is quicker if the mean withdrawal time from a random sample of 20 cash withdrawals is less than 1.7 minutes. Assume that, for the new system, the standard deviation is still 0.9 minutes, and the time spent still follows a normal distribution.

i) Calculate the probability of a Type I error. [4]

ii) If the mean withdrawal time under the new system is actually 1.5 minutes, calculate the probability of a Type II error. [4]

Cambridge International AS and A Level Mathematics 9709 Paper 7 Q5 June 2003

16. A survey taken last year showed that the mean number of computers per household in Branley was 1.66. This year a random sample of 50 households in Branley answered a questionnaire with the following results.

Number of computers	0	1	2	3	4	>4
Number of households	5	12	18	10	5	0

i) Calculate unbiased estimates for the population mean and variance of the number of computers per household in Branley this year. [3]

ii) Test at the 5% significance level whether the mean number of computers per household has changed since last year. [5]

iii) Explain whether it is possible that a Type I error may have been made in the test in part ii). [1]

iv) State what is meant by a Type II error in the context of the test in part ii), and give the set of values of the test statistic that could lead to a Type II error being made. [2]

Cambridge International AS and A Level Mathematics 9709 Paper 71 Q6 June 2012

17. People who diet can expect to lose an average of 3 kg in a month. In a book, the authors claim that people who follow a new diet will lose an average of more than 3 kg in a month. The weight losses of the 180 people in a random sample who had followed the new diet for a month were noted. The mean was 3.3 kg and the standard deviation was 2.8 kg.

i) Test the authors' claim at the 5% significance level, stating your null and alternative hypotheses. [5]

ii) State what is meant by a Type II error in words relating to the context of the test in part i). [2]

Cambridge International AS and A Level Mathematics 9709 Paper 7 Q4 June 2008

18. Past experience has shown that the heights of a certain variety of rose bush have been normally distributed with mean 85.0 cm. A new fertiliser is used and it is hoped that this will increase the heights. In order to test whether this is the case, a botanist records the heights, x cm, of a large random sample of n rose bushes and calculates that $\bar{x} = 85.7$ and $s = 4.8$, where $\bar{x}$ is the sample mean and s^2 is an unbiased estimate of the population variance. The botanist then carries out an appropriate hypothesis test.

i) The test statistic, z, has a value of 1.786 correct to 3 decimal places. Calculate the value of n. [3]

ii) Using this value of the test is statistic, carry out the test at the 5% significance level. [3]

Cambridge International AS and A Level Mathematics 9709 Paper 71 Q3 June 2011

Maths in real-life

A risky business

Companies would like to reduce the risk of various parts of their business as much as possible if they can do so at a reasonable cost. Financial products called *derivatives* are one of the ways of doing this and a simple example can illustrate why such markets have developed and are sustainable. Fuel is one of the biggest costs in an airline's business model, but the price on the open market fluctuates. Oil companies have a business model where the costs of extracting and refining oil into the fuel used by airlines are very substantial, and the price they are able to sell the fuel at is determined by market conditions after the company has spent that money in production. A *forward contract* where the airline commits to buying an amount of fuel from the oil company over the next year at a price which is fixed now can be beneficial to both companies in terms of their planning. It is likely that one or other of the companies would have been better off (this happens unless the price agreed is exactly what the market price ended up through the year) but they are both happy to have the security of removing the risk. The fair pricing of a contract like this involves mathematics and statistics as well as knowledge of what market conditions may influence price movements over the period of the contract.

Large financial institutions and wealthy individuals will put together a *portfolio* of investments – aiming for high returns with low *volatility* – meaning just that the returns are consistently high rather than very high sometimes and low on others.

This is a form of insurance for both companies, but it is a rather different structure than household or car insurances.

Diversification across sectors of the economy, and types of financial product, also helps to minimise the risks associated with a severe downturn in a particular sector. Increasingly sophisticated derivative products have been devised to attract investors – one of these is *statistical arbitrage* which illustrates the benefits of using large numbers of smaller investments (instead of a small number of large investments) to produce more stable returns. It exploits pricing inefficiencies between pairs of securities in the same sector. The principle is quite straightforward:

- Identify a pair of stocks, in the same sector, with similar company characteristics which you consider to have a relative difference in value.
- Buy forward the one you consider relatively undervalued, and sell forward the one you consider relatively overvalued.
- Any movement in the sector will apply to both, if your assessment was correct the relative difference in price change is likely to be in your favour.
- Can improve the risk / reward profile of this endeavour by applying it to a large number of smaller investments where the laws of large numbers kick in.

The graphs below illustrate the difference in the volatility of the three strategies used by A, B and C where A invests a large amount in each of a small number of trades, C invests a small amount in a large number of trades and B is in between. You can run the simulation for yourself from the website.

All three investors invest the same total amount, but C invests in a large number of trades with a small amount each; B invests 10 times as much in each trade as C but in only $\frac{1}{10}$ as many trades and A invests 100 times as much in each as C does but in $\frac{1}{100}$ of the trades.

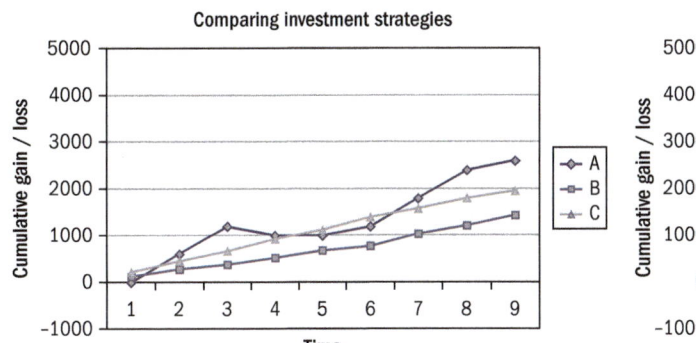

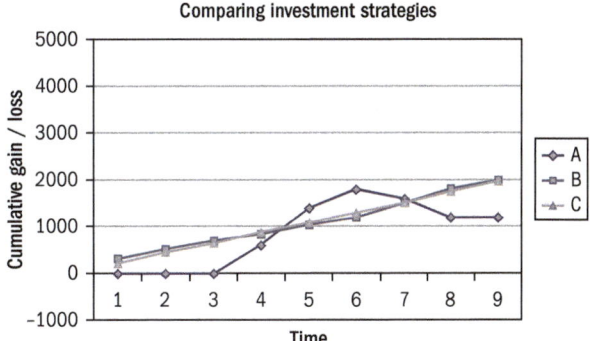

On average, the three strategies give the same return, but C is very stable – very little difference between simulations whereas A is extremely variable between trials.

List of formulae and tables of the normal distribution

If Z has a normal distribution with mean 0 and variance 1 then, for each volume of z, the table gives the values of $\Phi(z)$, where $\Phi(z) = P(Z \leq z)$.

For negative values of z use $\Phi(-z) = 1 - \Phi(z)$.

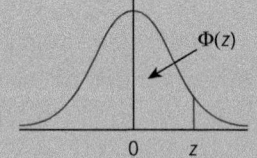

z	0	1	2	3	4	5	6	7	8	9	1	2	3	4	5	6	7	8	9
															ADD				
0.0	0.5000	0.5040	0.5080	0.5120	0.5160	0.5199	0.5239	0.5279	0.5319	0.5359	4	8	12	16	20	24	28	32	36
0.1	0.5398	0.5438	0.5478	0.5517	0.5557	0.5596	0.5636	0.5675	0.5714	0.5753	4	8	12	16	20	24	28	32	36
0.2	0.5793	0.5832	0.5871	0.5910	0.5948	0.5987	0.6026	0.6064	0.6103	0.6141	4	8	12	15	19	23	27	31	35
0.3	0.6179	0.6217	0.6255	0.6293	0.6331	0.6368	0.6406	0.6443	0.6480	0.6517	4	7	11	15	19	22	26	30	34
0.4	0.6554	0.6591	0.6628	0.6664	0.6700	0.6736	0.6772	0.6808	0.6844	0.6879	4	7	11	14	18	22	25	29	32
0.5	0.6915	0.6950	0.6985	0.7019	0.7054	0.7088	0.7123	0.7157	0.7190	0.7224	3	7	10	14	17	20	24	27	31
0.6	0.7257	0.7291	0.7324	0.7357	0.7389	0.7422	0.7454	0.7486	0.7517	0.7549	3	7	10	13	16	19	23	26	29
0.7	0.7580	0.7611	0.7642	0.7673	0.7704	0.7734	0.7764	0.7794	0.7823	0.7852	3	6	9	12	15	18	21	24	27
0.8	0.7881	0.7910	0.7939	0.7967	0.7995	0.8023	0.8051	0.8078	0.8106	0.8133	3	5	8	11	14	16	19	22	25
0.9	0.8159	0.8186	0.8212	0.8238	0.8264	0.8289	0.8315	0.8340	0.8365	0.8389	3	5	8	10	13	15	18	20	23
1.0	0.8413	0.8438	0.8461	0.8485	0.8508	0.8531	0.8554	0.8577	0.8599	0.8621	2	5	7	9	12	14	16	19	21
1.1	0.8643	0.8665	0.8686	0.8708	0.8729	0.8749	0.8770	0.8790	0.8810	0.8830	2	4	6	8	10	12	14	16	18
1.2	0.8849	0.8869	0.8888	0.8907	0.8925	0.8944	0.8962	0.8980	0.8997	0.9015	2	4	6	7	9	11	13	15	17
1.3	0.9032	0.9049	0.9066	0.9082	0.9099	0.9115	0.9131	0.9147	0.9162	0.9177	2	3	5	6	8	10	11	13	14
1.4	0.9192	0.9207	0.9222	0.9236	0.9251	0.9265	0.9279	0.9292	0.9306	0.9319	1	3	4	6	7	8	10	11	13
1.5	0.9332	0.9345	0.9357	0.9370	0.9382	0.9394	0.9406	0.9418	0.9429	0.9441	1	2	4	5	6	7	8	10	11
1.6	0.9452	0.9463	0.9474	0.9484	0.9495	0.9505	0.9515	0.9525	0.9535	0.9545	1	2	3	4	5	6	7	8	9
1.7	0.9554	0.9564	0.9573	0.9582	0.9591	0.9599	0.9608	0.9616	0.9625	0.9633	1	2	3	4	4	5	6	7	8
1.8	0.9641	0.9649	0.9656	0.9664	0.9671	0.9678	0.9686	0.9693	0.9699	0.9706	1	1	2	3	4	4	5	6	6
1.9	0.9713	0.9719	0.9726	0.9732	0.9738	0.9744	0.9750	0.9756	0.9761	0.9767	1	1	2	2	3	4	4	5	5
2.0	0.9772	0.9778	0.9783	0.9788	0.9793	0.9798	0.9803	0.9808	0.9812	0.9817	0	1	1	2	2	3	3	4	4
2.1	0.9821	0.9826	0.9830	0.9834	0.9838	0.9842	0.9846	0.9850	0.9854	0.9857	0	1	1	2	2	2	3	3	4
2.2	0.9861	0.9864	0.9868	0.9871	0.9875	0.9878	0.9881	0.9884	0.9887	0.9890	0	1	1	1	2	2	2	3	3
2.3	0.9893	0.9896	0.9898	0.9901	0.9904	0.9906	0.9909	0.9911	0.9913	0.9916	0	1	1	1	1	2	2	2	2
2.4	0.9918	0.9920	0.9922	0.9925	0.9927	0.9929	0.9931	0.9932	0.9934	0.9936	0	0	1	1	1	1	1	2	2
2.5	0.9938	0.9940	0.9941	0.9943	0.9945	0.9946	0.9948	0.9949	0.9951	0.9952	0	0	0	1	1	1	1	1	1
2.6	0.9953	0.9955	0.9956	0.9957	0.9959	0.9960	0.9961	0.9962	0.9963	0.9964	0	0	0	0	1	1	1	1	1
2.7	0.9965	0.9966	0.9967	0.9968	0.9969	0.9970	0.9971	0.9972	0.9973	0.9974	0	0	0	0	0	1	1	1	1
2.8	0.9974	0.9975	0.9976	0.9977	0.9977	0.9978	0.9979	0.9979	0.9980	0.9981	0	0	0	0	0	0	0	1	1
2.9	0.9981	0.9982	0.9982	0.9983	0.9984	0.9984	0.9985	0.9985	0.9986	0.9986	0	0	0	0	0	0	0	0	0

Critical values for the normal distribution

If Z has a normal distribution with mean 0 and variance 1 then, for each value of p, the table gives the value of z such that $P(Z \leq z) = p$.

p	0.75	0.90	0.95	0.975	0.99	0.995	0.9975	0.999	0.9995
z	0.674	1.282	1.645	1.960	2.326	2.576	2.807	3.090	3.291

Answers

The answers given here are concise. However, when answering exam-style questions, you should show as many steps in your working as possible.

Cambridge International Examinations bears no responsibility for the example answers to questions taken from its past paper questions which are contained in this publication.

1 The Poisson distribution

Skills check page 1

1. **a)** 0.0498 **b)** 0.1225

2. 0.101

Exercise 1.1 page 4

1. **i)** 0.271 **ii)** 0.271 **iii)** 0.180

2. **i)** 0.165 **ii)** 0.298 **iii)** 0.268

3. **i)** 0.124 **ii)** 0.174 **iii)** 0.116

4. **i)** 0.670 **ii)** 0.268 **iii)** 0.054

5. **i)** 0.269 **ii)** 0.104 **iii)** 0.016

6. 0.041; $P(X = 1) = 0.130$; $P(X = 2) = 0.209$
   i) $P(X = 2) = 0.209$
   ii) $P(X \leq 2) = 0.380$
   iii) $P(X \geq 2) = 0.829$

7. **a)** $P(X = 2) = 0.209$
   b) $P(X > 2) = 0.620$

8. $P(X = 0) = 0.273$; $P(X = 1) = 0.354$;
   $P(X = 2) = 0.230$
   a) $P(X = 2) = 0.230$
   b) $P(X < 2) = 0.627$

Exercise 1.2 page 6

1. $P(X = 0) = 0.061$; $P(X = 1) = 0.170$;
   $P(X = 2) = 0.238$
   a) $P(X = 2) = 0.238$
   b) $P(X > 2) = 0.531$

2. $P(X = 0) = 0.061$; $P(X = 1) = 0.170$;
   $P(X = 2) = 0.238$
   a) $P(X = 2) = 0.238$
   b) $P(X < 2) = 0.231$

3. **a)** $P(Y = 2) = 0.144$
   b) $P(X = 0) = 0.091$;
   $P(X = 1) = 0.218$;
   $P(X = 2) = 0.261$
   $P(X > 2) = 0.430$

4. **a)** $P(X = 1) = 0.164$
   b) $P(X \leq 2) = 0.833$
   c) $P(X = 6) = 0.146$

5. **a)** $P(X = 2) = 0.224$
   b) $P(X > 2) = 0.577$

Exercise 1.3 page 9

1. **a)** $P(X = 4) = \frac{2.5}{4} \times P(X = 3)$
   b) 0.134
   c) $P(X = 4) = 0.134$
   d) 2 is the mode

2. **a)** $P(X = 4) = \frac{4}{4} \times P(X = 3) = P(X = 3)$
   b) 3 and 4 are the modes

3. **a)** $\lambda = 4.8$
   b) 4 is the mode

Exercise 1.4 page 11

1. **i)** 3.2 **ii)** 3.2

2. 49; 7

3. **a)** $\mu = 3.6$; $\sigma = 1.90$
   b) 0.485
   c) 0.031
   d) 0

4. **a)** $\mu = 6$; $\sigma = 2.45$
   b) 0.394
   c) 0.043
   d) 0.017

5. Note that there is no 'right answer' to this question. In Q3, where λ is smaller the distribution is more heavily skewed (positively) and there are no possible outcomes more than 2 standard deviations below the mean, while in Q4 it is possible – although the likelihood is still small.

Exercise 1.5 page 16

1. **a)** yes: $\lambda = 3\left(=\dfrac{12}{4}\right)$

b) no – in a single lane the independence / randomness required for the Poisson is lost.

c) unlikely because it is so close to Christmas and the average rate is not likely to be the same as normal.

d) unlikely to be the same average rate over the whole 24 hour period

e) A&E is likely to be more busy at this time than usual because of the heavy traffic at the end of the working week.

2. **a)** and **b)** require the defects to occur independently of one another and at random. In part **a)** $\lambda = 1.6$ and in **b)** $\lambda = 4$

c) the customers need to arrive independently of one another and at random times – in 20 minutes use $\lambda = \dfrac{8}{3}$ (note that you might get two (or more) people entering the shop together but only one would actually be a customer.

3. **a)** 2.514; Var$(X) = 2.36$

b) the mean and variance are not all that similar so the Poisson is unlikely to be a good model for this situation.

4. **a)** **i)** 0.223　**ii)** 0.191　**iii)** 0.303

b) 0.193

c) probably not as the average rate in the middle of the night is unlikely to be the same.

5. **a)** 0.335　　**b)** 0.169　　　**c)** 0.222

Summary exercise 1 page 17

1. **a)** 0.247　　**b)** 0.821　　**c)** 0.425

2. **a)** **i)** 0.041　**ii)** 0.130　**iii)** 0.620

b) **i)** 3.2　　**ii)** 3.2　　**iii)** 1.79

c) $P(X = k + 1) = \dfrac{3.2}{k+1}P(X = k)$ which is multiplying by more than 1 when k is under 3, so the probabilities are increasing, but is then decreasing when k is 3 or greater so 3 has the largest probability

d) **i)** 0.603　**ii)** 0.609

3. **a)** **i)** 0.119　**ii)** 0.159

b) **i)** 6.4　　**ii)** 6.4　　**iii)** 2.53

c) 6

d) **i)** 0.119

ii) 0.684. In an hour on average 1.6 calls are received so use Po(1.6)

4. **a)** 0.258　　　**b)** 0.217

5. **a)** 0.258　　　**b)** 0.525

6. **a)** mean = 4. Standard dev = 2

b) 0.371　　　**c)** 0.547

7. **a)** $\bar{x} = 2.67$;　Variance $= 3.89$

b) No – the mean and variance are not close – and the pattern of results (where no accidents gives a second peak along with 4) is not typical of the single modal form of the standard Poisson.

8. **a)** P(one error) $= 0.129$

b) For 20 pages the total number of errors, X, has a Poisson distribution with mean 3

$P(X \le 2) = 0.423$

c) The most likely (equal likely) is 2 and 3 errors.

9. **a)** P(one error) $= 0.329$

b) P(two errors on double page) $= 0.217$

c) P(one error on each page | two errors on double page) $= 0.499$

10. **a)** $P(X = 2) = 0.245$

b) $P(X \le 4) = 0.863$

c) $0.057 < 0.1$

11. **a)** $E(X) = 2.5$;　Var$(X) = 2.5$;　$E(Y) = 5$;　Var$(Y) = 10$

b) $E(Y) \ne$ Var(Y) so Y can not have a Poisson distribution

12. **a)** 0.184　　　**b)** 0.125 (3 s.f.)

13. **a)** 0.469 (3 s.f.) **b)** 2.15 mins (3 s.f.)

2 Approximations involving the Poisson distribution

Skills check page 20

1. 0.881

2. 0.2585

Exercise 2.1 page 22

1. **a)** $X \sim B(60,0.05)$

    i)　$P(X = 0) = 0.046$　　**ii)**　$P(X = 1) = 0.145$

    iii) $P(X = 2) = 0.226$　**iv)**　$P(X > 2) = 0.583$

b)　$X \overset{approx}{\sim} \text{Po}(3)$

    i)　$P(X = 0) = 0.050$　　**ii)**　$P(X = 1) = 0.149$

    iii) $P(X = 2) = 0.224$　**iv)**　$P(X > 2) = 0.577$

2. a) $n > 50$ and $np < 5$ (approximately)

b) 0.000125

c) $X \overset{approx}{\sim} \text{Po}(4.5)$

$P(X < 3) = 0.174$

3. $X \overset{approx}{\sim} \text{Po}(4)$

$P(X < 5) = 0.629$

4. a) $X \sim \text{B}(130, 0.005)$

$P(X = 1) = 0.340$

b) $X \overset{approx}{\sim} \text{Po}(3)$

$P(X < 5) = 0.815$

5. a) $X \overset{approx}{\sim} \text{Po}(4)$

i) $P(X = 0) = 0.018$

ii) $P(X \geq 5) = 0.371$

b) Samples vary and you are not told how the sample was taken: a sample of 1000 is quite large but it would not mean you could be sure the level of support was higher than the party claimed. (You will consider issues like this in more depth in chapters 8 and 9 when you meet hypothesis tests.)

Exercise 2.2 page 25

1. a) $P(Y < 15.5)$

b) $P(Y > 22.5)$

c) $P(Y \leq 17.5)$

d) $P(44.5 \leq Y < 61.5)$

2. a) Yes **b)** No **c)** No

3. a) 0.142 **b)** 0.984 **c)** 0.916

d) 0.163 **e)** 0.6103

4. a) i) 0.424

ii) 0.440

b) i) 0.00016

ii) 0.0377%

5. If λ is greater than approximately 15 then the $N(\lambda, \lambda)$ can be used with the continuity correction to approximate the $\text{Po}(\lambda)$ distribution.

6. a) 0.185 **b)** 0.123

7. a) 0.251 **b)** 0.614

8. 0.182

Summary exercise 2 page 26

1. a) i) 0.223

ii) 0.337

iii) 0.255

iv) 0.190

Note that the answer to **iv)** is 0.190 correct to 3 dp although using answers to parts **i)** to **iii** to 3 dp will give 0.189

b) $X \sim \text{B}(60, 0.025) \overset{approx}{\sim} \text{Po}(1.5)$

ii) $P(X = 0) = 0.335$

iv) $P(X = 2) = 0.191$

2. a) $X \sim \text{B}\left(180, \frac{1}{150}\right)$ $P(X = 1) = 0.362$

b) $X \sim \text{B}\left(500, \frac{1}{150}\right) \overset{approx}{\sim} \text{Po}\left(\frac{10}{3}\right)$ $P(X < 5) = 0.756$

3. a) $X \sim \text{B}\left(7500, \frac{1}{200}\right) \overset{approx}{\sim} \text{Po}(3.75)$ $P(X > 3) = 0.516$

b) Smallest possible n is 9211

4. a) They need to arrive independently and at random (either condition will be enough)

b) using $\text{Po}(5)$: $P(X = 4) = 0.175$

c) using $\text{Po}(2.5)$: $P(Y < 3) = 0.544$

d) using $\text{Po}(30) \overset{approx}{\sim} \text{N}(30, 30)$: $P(W < 25) = 0.158$

5. a) i) 0.713 (0.7127)

ii) 0.360

b) 0.219

6. $P(X > 2) = 0.121$

7. 0.515

8. a) $\text{B}(40{,}000, 0.0001)$

b) $\text{Po}(4)$

$n = 40{,}000 > 50$, $np = 4 < 5$

c) 0.567

9. a) $\text{B}(520, 0.008)$

$\text{Po}(4.16)$

$np = 4.16 < 5$

b) i) 0.597 (3 s.f.)

ii) Smallest n is 4

10. 0.0328(3 s.f.)

3 Linear combination of random variables

Skills check page 28

1. $E(X) = 2.4$; $Var(X) = 6.64$

Exercise 3.1 page 31

1. a) $E(X) = 18.4$ $Var(X) = 7.6$

   b) $E(X) = -13.1$ $Var(X) = 17.1$

   c) $E(X) = 8.7$ $Var(X) = 7.6$

   d) $E(X) = 39.9$ $Var(X) = 93.1$

2. a) i) $E(X) = 3.1$
      $E(X^2) = 10.9$
      $Var(X) = 1.29$

   ii) $E(5X - 2) = 13.5$; $Var(5X - 2) = 32.25$

   b) i) $E(Y) = 0.1$
      $E(Y^2) = 1.3$
      $Var(Y) = 1.29$

   ii) $E(5Y - 2) = 5 - 1.5$; $Var(5Y - 2) = 32.25$

   iii) $E(X) - 3 = E(Y)$; $Var(X) = Var(Y)$

   c) i) $E(V) = 8.25$
      $E(V^2) = 72.875$
      $Var(V) = 4.8125$

   ii) $E(4 + 3V) = 24.75$; $Var(4 + 3V) = 43.3125$

   d) i) $E(W) = -0.756$, $Var(W) = 4.8125$

   ii) $E(4 + 3W) = 1.75$; $Var(4.3W) = 43.3125$

   iii) $E(W) = E(V) - 9$; $Var(W) = Var(V)$

3. i) $E(X) = -0.2$
   $E(X^2) = 1.4$
   $Var(X) = 1.36$

   ii) $E(7 - 2X) = 5.44$

4. a) $E(X) = 12.25$
   $E(X^2) = 178.75$
   $Var(X) = 28.6875$

   b) 0.6

   c) $E(0.9X) = 11.025$; $Var(0.9X) = 23.2$

   d) i) 11.25; 28.6875

   ii) Buying 4 x \$5 vouchers would get \$4 off instead of just a \$1 reduction

5. a) $P(X = 8) = 0.3$ b) $E(X) = 8$

   c) 65
      $Var(X) = 1$

   d) $E(5 - 2X) = -13$

   i) 11.25; 28.6875

   ii) Buying $4 \times \$5$ vouchers would get \$4 off instead of just a \$1 reduction

   e) $Var(5 - 2X) = 4$

6. a) $E(X) = 9.05$

   b) $b = 0.25$; $a = 0.15$

   c) $Var(X) = 2.0475$

   d) $Var(5 - 3X) = 18.4275$

7. a)

x	1	2	3	4	5
$P(X = x)$	$\frac{9}{25}$	$\frac{7}{25}$	$\frac{5}{25}$	$\frac{3}{25}$	$\frac{1}{25}$

   b) $P(2 < X < 5) = \frac{8}{25}$ c) $E(X) = 2.2$

   d) $Var(X) = 1.36$ e) $Var(2 - 3X) = 12.24$

8. a) $E(X) = 17.85$ b) $Var(X) = 26.7275$

   c) $Var(20 - 3X) = 240.5475$

9. a) $k = \frac{1}{28}$

   b) $E(X) = 5$
      $E(X^2) = 28$
      $Var(X) = 3$

   c) Makes a loss if no more than 2 options are successful, so probability $= \frac{3}{28}$

   d) $E(P) = 3500$;
      $Var(P) = 6750000$

   This would be a good investment portfolio, spread over a number of individual investments, with a low probability of making a loss and with an average return of 87.5% of the capital at risk.

10. a) $E(X) = 2.8$ Standard deviation $= 1.29$

    b) $E(Y) = 33$;   Standard deviation $= 12.9$

Exercise 3.2 page 37

1. i) $E(X + Y) = 16$ $Var(X + Y) = 7$

   ii) $E(X - Y) = -4$ $Var(X - Y) = 7$

2. **i)** $E(X + Y) = 30$ $Var(X + Y) = 6$
   ii) $E(X - Y) = 0$ $Var(X - Y) = 6$

3. **i)** $E(X + Y) = 3$ $Var(X + Y) = 3.5$
   ii) $E(X - Y) = 11$ $Var(X - Y) = 3.5$

4. $E(M) = 70.275$
   standard deviation of $M = 11.1$

5. **i)** $E(C) = 105$
      standard deviation of $M = 15$
   ii) It is likely that many customers will usually
      use only one of the two branches but that
      there will be some customers who use both
      – depending on their movements on the
      day they need to visit the bank so assuming
      independence of the number of customers is
      likely not to be true but could still be a useful
      first approximation.

6. **i)** $E(C) = \frac{1}{2} \Rightarrow Var(C) = \frac{1}{4}$
   ii) $E(X) = 2.5$
      $Var(X) = \frac{47}{12}$

Exercise 3.3 page 39

1. **i)** $E\left(\sum_{i=1}^{6} X_i\right) = 90$
      $Var\left(\sum_{i=1}^{6} X_i\right) = 24$
   ii) $E(\bar{X}_6) = \frac{2}{3}$

2. **i)** $E\left(\sum_{i=1}^{20} X_i\right) = 0$
      $Var\left(\sum_{i=1}^{20} X_i\right) = 400$
   ii) $E(\bar{X}_{20}) = 1$ $Var(\bar{X}_{20}) = 1$

3. **i)** $E\left(\sum_{i=1}^{50} X_i\right) = -250$
      $Var\left(\sum_{i=1}^{50} X_i\right) = 1250$
   ii) $E(\bar{X}_{50}) = -5$
      $Var(\bar{X}_{50}) = 0.5$

4. **i)** $E\left(\sum_{i=1}^{10} X_i\right) = 352$
      $Var\left(\sum_{i=1}^{10} X_i\right) = 261$
   ii) $E(\bar{X}_{10}) = 35.2$
      $Var(\bar{X}_{10}) = 2.61$

5. The sample size needed is 6

6. The sample size needed is 40

Summary exercise 3 page 41

1. **a)** $E(2X + 3) = 27.2$
      $Var(2X + 3) = 9.6$
   b) $E(5 - 3X) = -31.3$
      $Var(5 - 3X) = 21.6$
   c) $E(X - 3) = 9.1$
      $Var(X - 3) = 2.4$
   d) $E(9X) = 108.9$
      $Var(9X) = 194.4$

2. **i)** $E(X) = 14$
      $Var(X) = 1.2$
   ii) $E(4X - 3) = 53$;
      $Var(4X - 3) = 19.2$

3. **a)** $\Sigma p = 1$
      $E(X) = 9.2 \Rightarrow 1.4 + 8a + 2.7 + 1 + 11b = 9.2$
   b) $b = 0.3$; $a = 0.1$
   c) $Var(X) = 2.16$
   d) $Var(3 - 2X) = 8.64$

4. **a)**

x	1	2	3	4
$P(X = x)$	$\frac{7}{16}$	$\frac{5}{16}$	$\frac{3}{16}$	$\frac{1}{16}$

   b) $P(2 \leq X \leq 4) = \frac{9}{16}$
   c) $E(X) = \frac{15}{8}$
   d) $E(X^2) = \frac{35}{8}$
      $Var(X) = 0.859(375)$
   e) $Var(2 - 3X) = 7.73$

5. **i)** $E(X + Y) = 17$
      $Var(X + Y) = 6$
   ii) $E(X - Y) = 3$
      $Var(X - Y) = 6$

6. $E(M) = 59.2$

standard deviation of $M = 5.60$

Note – it may be a little surprising that this is lower than either standard deviation which goes to make it up – but it is because you are taking the average of the two component marks.

7. a) $k = \frac{1}{30}$

b) $E(X) = \frac{10}{3}$

$E(X^2) = 11.8$

c) Var $(X) = 0.689$

d) 3.67; 188.8

8. a) $\Sigma p = 1$

$E(X) = 4.4$

b) $b = 0.4;\quad a = 0.1$

c) $E(3X - 4) = 9.2$

d) $\text{Var}(X) = 9.24$

e) $\text{Var}(3X - 4) = 83.16$

9. $a = 0.8 \quad b = 18.4$

10. i) 2.5; 1.25 **ii)** 5; 5

11. i) −2 **ii)** 0.972

Review exercise A page 44

1. $a = 0.8, b = -1$

2. i) 0.209 **ii)** 0.219 **iii)** $n = 6$

3. 0.428

4. i) 0.144 **ii)** 0.819 **iii)** $t = 2.88$

5. i) 0.269 **ii)** $n = 6908$

6. i) 0.918

ii) a) $k = 2.48$ **b)** 0.915

7. i) 5.2 **ii)** Var ≠ mean

8. i) 105 cents

ii) 257 cents (3 s.f.)

9. i) 0.122 **ii)** 0.532 **iii)** 0.0135

iv) 0.229

10. i) 0.143 **ii)** 0.118 **iii)** 0.0316

11. i) 0.815

ii) Least value of $n = 58$

12. 12.1 (3 s.f.)

4 Linear combination of Poisson and Normal variables

Skills check page 48

1. 0.257

2. 0.275

Exercise 4.1 page 51

1. a) 0.040 **b)** 0.120 **c)** 0.265

2. a) i) 0.202 **ii)** 0.857 **iii)** 0.026

b) i) $S \sim \text{Po}(1.6);\quad M \sim \text{Po}(5.4) \quad \Rightarrow T \sim \text{Po}(7)$

ii) 0.827

3. a) i) 0.169 **ii)** 0.793

4 only part **iii** is a Poisson

5. 0.132

6. 0.032

Exercise 4.2 page 56

1. all of them have a Normal distribution

2. i) $2X + 3Y \sim \text{N}(970, 350)$

ii) 0.0544

3. i) $5X + 4Y \sim \text{N}(450, 306.1)$

ii) 0.716

4. i) $4A - 3B \sim \text{N}(57, 130)$

ii) 0.570

5. i) $C \sim \text{N}(3444.8, 5.8^2 \times 18.2^2)$

ii) 0.301

6. i) 0.9895

ii) $P\left(\sum_{i=1}^{4} W_i < 3000\right) = 0.0787$

7. i) 0.868 **ii)** 1.000

Summary exercise 4 page 57

1. a) i) 0.0273 **ii)** 0.983 **iii)** 0.1005

b) i) $Y \sim \text{Po}(3.9)$ **ii)** 0.352

2. i) 0.187 **ii)** 0.00826

3. i) $E(C) = 6676$

st dev of $C = 532.48 \Rightarrow C \sim \text{N}(6676, 532.48^2)$

ii) $\Phi(0.330) = 0.629$

4. i) $1 - \Phi(0.386) = 0.350$

   ii) $\Phi(0.877) = 0.810$

5. $\Phi(0.069) = 0.5275$

6. i) 1.4   **ii)** 0.0596   **iii)** 0.250

   iv) $P(X \geq 1) = 0.909$

   v) $P(T > 3) = 0.658$

7. $1 - \Phi(1.076) = 0.141$

8. $1 - \Phi(0.338) = 0.368$

9. i) 0.326   **ii)** $n = 36$

10. 0.0147 (3 s.f.)

11. i) 0.889 (3 s.f.) **ii)** 0.176 (3 s.f.)

12. i) 0.387   **ii)** Mean: 10; variance: 5.78

   iii) 0.647

13. i) 0.875   **ii)** 0.0285 to 0.0286

   iii) 0.246 (3 s.f.)

14. i) 9.5; 0.524   **ii)** 0.972

5 Continuous random variables

Skills check page 60

1. $k = 2$

2. 15

Exercise 5.1 page 62

1. **a**, **c** and **d** are continuous (although they will be reported to specified accuracies) while **b** and **e** are counts – discrete variables, although **b** would be much harder to count than **e** would be.

2.

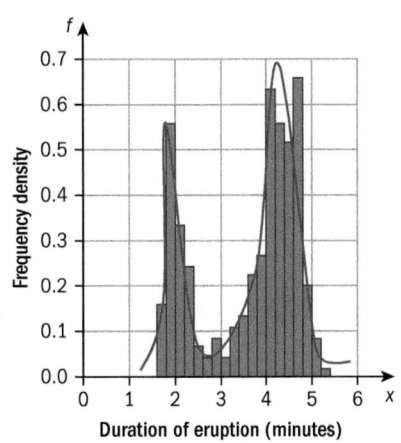

3.

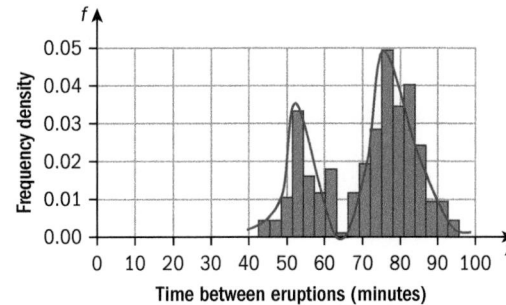

Time between eruptions (minutes)

4.

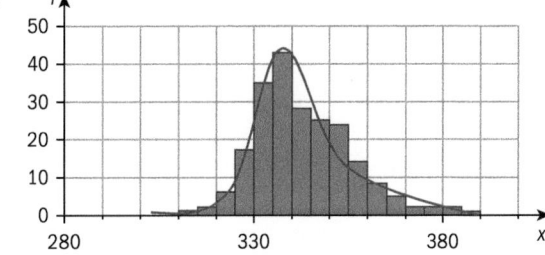

Exercise 5.2 page 66

1. i) yes – the area of the triangle is 1 and all non-negative

   ii) no – the area is 1 but part of the pdf is negative

   iii) no – the area of the triangle is 1.5

   iv) no – each of the grid squares in 0.25 square units and there are 4 squares fully shaded along with a lot of large fractions of other squares so total is > 1.

2. a) i) 1   **ii)** 0.5

   b) i) 0.25   **ii)** 0.5625

3. a) $k = 3$ **b)** $k = \frac{2}{9}$ **c)** $k = \frac{1}{12}$

   d) $k = 3$ **e)** $k = \frac{3}{2}$

4. a) $\frac{1}{4}$ **b)** $\frac{3}{16}$

   c) $k = \frac{2}{5}$ $P(X > 1) = \frac{11}{20}$

   d) i) $P(X < 1) = \frac{1}{8}$

      ii) $P(X = 1.5) = 0$ (the probability for any single point is zero)

   e) $k = \frac{3}{8}$

    $P(X < 2.5) = \frac{9}{16}$

5. **a)** $f(x) = \begin{cases} -\dfrac{2}{9}x + \dfrac{2}{3} & 0 \le x \le 3 \\ 0 & \text{elsewhere} \end{cases}$

b) $f(x) = \begin{cases} x + \dfrac{1}{2} & 0 \le x \le 1 \\ 0 & \text{elsewhere} \end{cases}$

6. **a)** **i)** $P(X > 2) = \dfrac{1}{9}$

ii) $P(X < 0.5) = \dfrac{11}{36}$

b) **i)** $P(X > 0.2) = 0.88$

ii) $P(X < 1) = 1$ since the probability of 1 itself is zero

Exercise 5.3 page 73

1. $\mu = 4.52$ $E(X^2) = \dfrac{41}{2}$

$\text{Var}(X) = 0.0830$

2. **a)** $k = \dfrac{2}{9}$

b) $\mu = 2$ $E(X^2) = \dfrac{9}{2}$

$\sigma^2 = 0.5$

c) $P(X > \mu) = \dfrac{5}{9}$

$\sigma = 0.707 \Rightarrow P(X > \mu + \sigma) = 0.186$

3. **a)** $k = \dfrac{4}{625}$

b) $\mu = 4$

$E(X^2) = \dfrac{50}{3}$

$\sigma^2 = \dfrac{2}{3}$

c) **i)** $P(X > \mu) = \dfrac{369}{625}$

ii) $\sigma = 0.8165 \Rightarrow P(X > \mu - 2\sigma) = 0.950$

4. **a)** $k = 5$

b) $\mu = \dfrac{5}{6}$

$E(X^2) = \dfrac{5}{7}$

$\sigma^2 = 0.01984....$

c) **i)** $P(X < \mu) = 0.402$

ii) $\sigma = 0.1408...$

$\Rightarrow P(|X - \mu| < \sigma) = 0.718$

5. **a)** $E(X) = 6$

b) $k = 12$

c) $E(X^2) = 36.15$

$\sigma^2 = 0.15$

6. **a)** **i)** If the bus arrives regularly but Gupta arrives at random times then the uniform distribution over the 15 minute interval is a sensible model.

ii) $\mu = 7.5$ (by symmetry)

$E(X^2) = 75$ $\text{Var}(X) = 18.75$

b) **i)** $k = \dfrac{2}{9}$

ii) $\mu = 1$ $E(Y^2) = 1.5$

$\text{Var}(Y) = 0.5$

7. **b)** standard deviation $= 1.22$

c) $P(\text{loss}) = \dfrac{7}{125}$

$P(\text{better return}) = 0.6875$

8. **b)** $\sigma = 0.6$

Exercise 5.4 page 76

1. **b)** **i)** mode – there are two – at 0 and 2

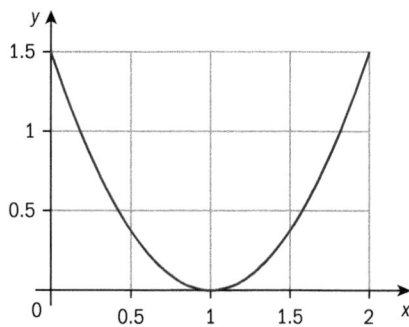

ii) symmetry of the distribution (quadratic function) says that the median is 1

2. **a)** **i)** if α is the median then the trapezium to the left of α under the pdf has width $(\alpha - 1)$ and vertical sides height $\dfrac{1}{12}$ & $\dfrac{\alpha}{12}$ and is to have area $\dfrac{1}{2}$

So $\left(\dfrac{\alpha + 1}{24}\right)(\alpha - 1) = \dfrac{1}{2} \Rightarrow \alpha = \sqrt{13}$

ii) the pdf is strictly montonic so the mode is at the end of interval (at 5)

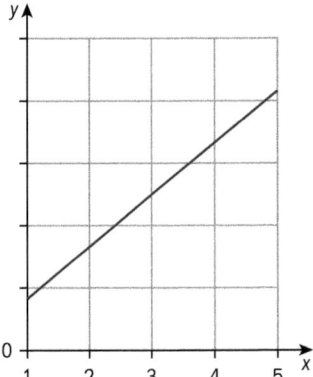

b) i) similar reasoning to part **a)** gives median as $\alpha = \sqrt{2}$

ii) mode is at 2

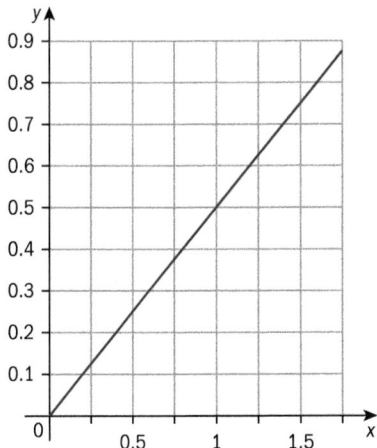

c) need the value of α for which $\alpha = 1.59$

ii) the mode of X is 2.

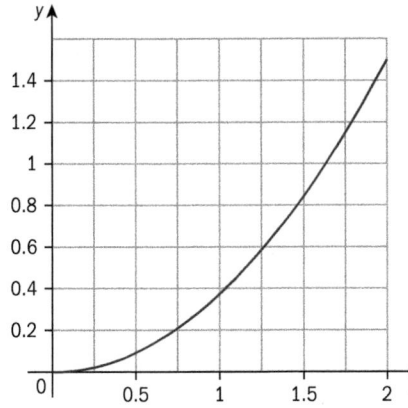

3. i) $k = \dfrac{3}{392}$

ii) The mode is at the maximum of the pdf – from the sketching process you can deduce it is at 5, or you could use differentiation to identify where the gradient is zero.

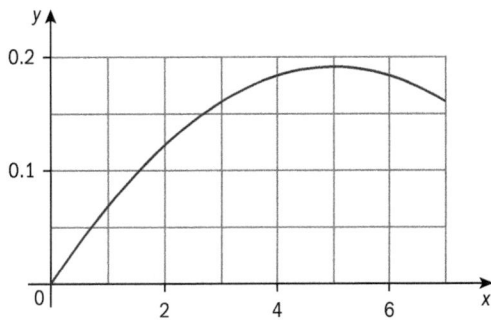

Summary exercise 5 page 77

1. a) $k = \dfrac{3}{13}$

b) $P(0.5 \le x \le 1) = 0.53$

c) $E(X) = 0.519$

$E(X^2) = \dfrac{23}{65}$ $Var(X) = 0.0842$

2. a) $C = \dfrac{1}{12}$

b) $P(1 \le x \le 2) = 0.33$ (2 d.p.)

c) $E(X) = \dfrac{15}{8}$

$E(X^2) = \dfrac{33}{8}$

$Var(X) = 0.609$

3. a) $P(1 \le x \le 2) = \dfrac{1}{4}$

(or argue geometrically from area of triangle)

b) $E(X) = \dfrac{2}{3}$

$E(X^2) = \dfrac{2}{3}$ $Var(X) = 0.222$

c) $P(X < \mu) = \dfrac{5}{9}$

4. i) $P(2.7 \le x \le 2.9) = 0.5$

(or argue geometrically from symmetry of pdf)

ii) $P(2.9 \le x \le 3) = 0.34375$

5 **a)** $P(x > 0.5) = 0.0625$

b) $E(X) = \frac{1}{5}$

$E(X^2) = \frac{1}{15}$

$Var(X) = 0.0267$

6. **a)** $k = \sqrt{2}$

b) 0.449

c) Q3 = 0.559

d) 0.0879

7. **b)** 2.66 hours

c) $m = 2.73$ hours

d) 0.0243

8. **b)** $x = 4.36$ (3 s.f.)

9. **a)** $t = 8.41$ mins (3 s.f.)

b) 6.33 (3 s.f.); 2.09 (3 s.f.)

10. **a)** **i)** X or 5   **ii)** V or 3

b) 0.75

c) **ii)** $a = 5, n = 4$

6 Sampling

Skills check page 80

1. 1,1; 1,2; 1,3; 1,4; 2,1; 2,2; 2,3; 2,4; 3,1; 3,2; 3,3; 3,4; 4,1; 4,2; 4,3; 4,4

Exercise 6.1 page 83

Note that in this chapter, many answers are only indicative of the range of possible responses.

1. **a)** **i)** a population is the complete group of people (or observations) of interest in a statistical investigation.

ii) A census is when information is gathered about all members of the population.

b) no single answer here – your example should show clearly why it is biased – for example, asking about health issues outside a fitness centre.

Labeling in the proof should be **i)** and **ii)** in part **b)**

2. **a)** **i)** a list in which all members of the population appear.

ii) a sampling unit can be an individual, or a household, or the result of an experiment (e.g. how many heads occur in 3 tosses of a coin)

b) **i)** all adults in the UK, could use the latest electoral register, or census files if it is close to the time at which the census was last done (only done every 10 years)

ii) the adults in a country changes every day – people move in and out; some die; others become adults (on their 18th birthday in the UK); and any such register will never have everyone on it that should be there, so even on the day the information was collected it would be 100% accurate.

3. Each member of the population has an equal chance of being selected, and all possible combinations are equally likely.

4. **a)** all the customers of that bank

b) use the account records held by the bank

c) an individual customer may hold more than one account, possibly even in more than one name (a woman might have accounts in her maiden name and married name), but more than one person can have the same name so just removing duplicate names may actually remove customers.

Exercise 6.2 page 85

1. **a)** unlikely to get a representative sample with this method – any reason which identifies a group likely to be over, or under-represented in the sample is a good answer, e.g. people who have more time are more likely to fill out the questionnaires

b) unless the route is fully booked the offer costs the airline almost nothing, and getting people to fill out the questionnaire engages their attention to consider the possibility of flying on that route at some point (and paying for a ticket!).

2. anything where the sample is just done by who happens to be in a particular place at a particular time is the easiest way to identify a convenience sample, e.g. outside a supermarket / train station / school ....

3. a) if they were disposable batteries, then a census would not be possible as there would be no batteries to sell. For rechargeable batteries it would be possible, but very expensive to test them all.

b) testing all 50 in one box means that the box can be removed where taking only one or two from a box until you have 50 batteries means a number of boxes are affected and have to be removed until they have 50 batteries again. That problem could be avoided by taking a random sample of 50 at a point in the process before they get packed into the boxes.

4. a) a census would be very time consuming and expensive.

b) they would be more concerned with keeping their important customers happy so they may wish to deliberately set up a sampling method designed to focus on particular types of account.

Exercise 6.3 page 89

1. methods **a)** and **e)** will give a random sample – in **e)** it does not matter that the sampling frame has been constructed in a non-random manner because the method of choosing the members of the sample is random.

2. the following samples for **a, b** are based on using the two digit numbers in the table, ignoring 00 and 91–99 and any repeats until 10 different numbers are obtained:

a) 52, 44, 17, 71, 20, 63, 47, 88, 22, 02

b) 52, 54, 07, 08, 43, 49, 24, 73, 67, 64

c) for a small sample you could just ignore 00 and 43–99 but it is very wasteful of the random numbers – so you could take 01–42 and also 51–92 (after subtracting 50 from anything in this range)

just using 01–42 gives: 17, 20, 22, 02, 18 using the subtraction method gives: 02, 17, 31, 20, 13 – note that rather than taking the next block of 42 numbers immediately after the first, the arithmetic is much much simpler to subtract 50 – it means that 00, 43–50 and 93–99 are not used which looks a little strange, but it is much easier.

d) as long as you specify what you intend to do before looking at the table of numbers there is no reason why it has to start at the top left.

Just using 01–38 gives: 07, 20, 22, 16, 11, 08

using 01–38 and 41–78 (after subtracting 40 in this range) gives 07, 27, 36, 20, 15, 22

3. the actual samples depends on the numbers generated by your calculator. But it should give you an insight into the amount of variability in the answers taking small samples – everyone owned a mobile phone so there is no variability there and the favourite type of film does have, but it is not numerical so there is no easy way to summarise the variability. The times are all relatively small values with a number of repeats so the averages of your sample times are likely to be closer together than the average number of DVDs owned.

4 and 5: again, the actual samples depends on the numbers generated by your calculator. These practical activities are designed both to give you a little bit of experience in the process of generating samples (pretty tedious to do lots of it …) and the opportunity to get some understanding of just how much variability results from taking random samples.

Exercise 6.4 page 95

1. A statistic is just a function of (some or all of) the set of observations collected, so **a** to **d** are all statistics but **e** is not as it involves parameters.

2. a) $\sum_{i=1}^{20} X_i \sim B\left(20, \frac{1}{3}\right)$ **b)** $P\left(\sum_{i=1}^{20} X_i = 2\right) = 0.0143$

c) $n = 20,\ p = \frac{1}{3} \Rightarrow E(X) = \frac{20}{3};\ Var(X) = \frac{40}{9}$

3. a) $\sum_{i=1}^{10} X_i \sim B\left(10, \frac{3}{4}\right)$ **b)** $P\left(\sum_{i=1}^{10} X_i \leq 7\right) = 0.474$

c) $n = 10,\ p = \frac{3}{4} \Rightarrow E(X) = 7.5;\quad Var(X) = 1.875$

4. a) $\mu = 2.6$

b) (1, 1, 1) (1, 1, 3) (1, 3, 1) (1, 3, 3) (3, 1, 1) (3, 1, 3) (3, 3, 1) (3, 3, 3)

These 8 possible samples have means:

$1, \frac{5}{3}, \frac{5}{3}, \frac{7}{3}, \frac{5}{3}, \frac{7}{3}, \frac{7}{3}$ and 3 and the modes are

1, 1, 1, 3, 1, 3, 3, 3 and the medians are 1, 1, 1, 3, 1, 3, 3, 3

c)

Sample mean	1	$\frac{5}{3}$	$\frac{7}{3}$	3
probability	$\frac{1}{8}$	$\frac{3}{8}$	$\frac{3}{8}$	$\frac{1}{8}$

d)

Sample mode	1	3
probability	$\frac{1}{2}$	$\frac{1}{2}$

e)

Sample median	1	3
probability	$\frac{1}{2}$	$\frac{1}{2}$

5. a) $\mu = 3.8$

b) (1, 1) (1, 3) (1, 6) (3, 1) (3, 3) (3, 6)
(6, 1) (6, 3) (6, 6)

These 9 possible samples have means: 1, 2, 3.5, 2, 3, 4.5, 3.5, 4.5, 6 and the medians are also 1, 2, 3.5, 2, 3, 4.5, 3.5, 4.5, 6 (because there are only two observations in a sample the median will be the same as the mean)

c)

Sample mean	1	2	3	3.5	4.5	9
probability	$\frac{1}{9}$	$\frac{2}{9}$	$\frac{1}{9}$	$\frac{2}{9}$	$\frac{2}{9}$	$\frac{1}{9}$

d)

Sample median	1	2	3	3.5	4.5	9
probability	$\frac{1}{9}$	$\frac{2}{9}$	$\frac{1}{9}$	$\frac{2}{9}$	$\frac{2}{9}$	$\frac{1}{9}$

Exercise 6.5 page 98

1. a) $E(\bar{X}_{10}) = 5$; $Var(\bar{X}_{10}) = 0.4$

b) $E(\bar{X}_{10}) = 26.3$; $Var(\bar{X}_{10}) = 0.78$

c) $E(\bar{X}_{10}) = 2$; $Var(\bar{X}_{10}) = 0.1$

d) $E(X) = 0$

2. i) $Var(\bar{X}_{15}) = 1.54$

ii) The minimum sample size is 24

3. i) $E(X) = 5$; $E(X^2) = 33.5$ $\Rightarrow Var(X) = 8.5$

ii) $\Rightarrow E(\bar{X}_{12}) = 5$; $Var(\bar{X}_{12}) = 0.708$

Exercise 6.6 page 100

1. i) $\bar{X}_6 \sim N\left(600, \frac{7.2^2}{6}\right)$ **ii)** $P(\bar{X} < 597) = 0.1536$

2. i) $\bar{X}_{10} \sim N\left(352, \frac{4.5^2}{10}\right)$ $P(\bar{X}_{10} \geq 350) = 0.920$

ii) The smallest sample size is 28.

3. i) The smallest sample size is 44.   **ii)** 0.493

4. i) 0.139   **ii)** 0.167

5. 0.141

Exercise 6.7 page 104

1. 0.0941   **2.** 0.972

3. i) 0.931

ii) the sampling distribution for the mean of a large sample uses the Central Limit Theorem to allow calculation of probabilities using the Normal distribution as an approximation, but for small samples the distribution is not known and probabilities can not be calculated.

4. $X \sim Po(7) \Rightarrow E(X) = Var(X) = 7$

i) CLT $\Rightarrow \bar{X}_{72} \overset{approx}{\sim} N\left(7, \frac{7}{72}\right)$

ii) $\Phi(1.604) = 0.946$

5. $X \sim B(10, 0.3) \Rightarrow E(X) = 3, Var(X) = 2.1$

i) CLT $\Rightarrow \bar{X}_{80} \overset{approx}{\sim} N\left(3, \frac{2.1}{80}\right)$

ii) $1 - \Phi(1.543) = 0.0614$

Summary exercise 6 page 107

1. While there are three outcomes (home win, draw and away win) these are not equi-probable outcomes (there tend to be more home wins in most league formats).

2. a) a random sample is one in which each member of the population has an equal chance of being selected, and all possible combinations are equally likely.

b) any reason which identifies a group likely to be over, or under-represented in the sample is a good answer.

3. It would be very expensive to test every battery for its usable life before it was sold – and they would need to be charged again.

4. Your answer will depend on the particular random sample you have taken, but many sample means will lie in, or close to, the interval 55 – 60.

5. **i)** AB AC AD BA BC BD CA
CB CD DA DB DC

ii) If A, B are 50 cent coins, C is the 20 cent coin and D is the 10 cent coin then the mean values of these samples (in the same order) are
50 35 30 50 35 30 35 35 15 30
30 15

Sample mean	15	30	35	50
probability	$\frac{2}{12}$	$\frac{4}{12}$	$\frac{4}{12}$	$\frac{2}{12}$

6. $E(X) = 22.4$ $Var(X) = 7.9$

i) $Var\left(\overline{X}_{20}\right) = 0.395$

ii) The smallest sample size is 8

7. **i)** $\overline{F}_6 \sim N\left(505, \frac{6.1^2}{6}\right)$

ii) $1 - \Phi(2.008) = 0.0223$

8. **i)** $\overline{M}_{10} \sim N\left(252, \frac{3.5^2}{10}\right)$

$\Phi(1.807) = 0.965$

ii) The smallest sample size is 21.

9. **a)** P(one of each = 0.5), P(HH) = 0.25, P(TT) = 0.25; One of each is more likely

b) Choose Charlie only if H then T; throw again if T then H

10. **a)** $\mu; \frac{\sigma^2}{n}$

b) **i)** normal

ii) unknown, or normal if the population is normal

Review exercise B page 109

1. **a)** Not all totals have the same probability

b) e.g. calculator, paper, random number tables

2. **b)** 0.227   **c)** $\frac{55}{36}$

3. **a)** $\frac{1}{12}$ **b)** Arrive every 12 minutes

4. **a)** $z_1 = 0.5587, z_2 = 1.117; 0.580$

b) CLT application

5. **a)** 0.0349   **b)** 99.5; 113.4   **c)** 0.879

6. **a)** 0.294 (3 s.f.)   **b)** 5

c) 0.0753 to 0.0754

7. **a)** All stayed between 1 and 9 hours

c) 3 hours

d) $x > 1.39$ hours   **e)** 0.774

8. **a)** 0.324   **b)** 0.994

9. **b)** $\frac{72}{125}$ **c)** $\frac{8}{3}$

10. **a)** 0.0625   **b)** $\frac{2}{3}$ **c)** $m = 0.586$

11. **a)** g: Area $\neq 1$ or > 1; h: pdf cannot be negative

b) **ii)** 0.0337 (3 s.f.)

12. 0.713

13. **a)** $6; \frac{2.6}{120}$ **b)** 0.282

14. **a)** $\overline{X} \sim N\left(48.8, \frac{15.6^2}{5}\right)$ **b)** 0.568

15. **a)** e.g. names in a hat   **b)** 16.6; 27.1

c) More

d) Pocket money of all pupils in Jenny's year

16. **a)** Not representative of whole population

b) People who travel to work on this train

c) 6.17; 0.657

17. **b)** $a = \frac{1}{9}$

c) $m < 0.5$; area below $x = 0.5$ is greater than 0.5

7 Estimation

Skills check page 116

1. mean = 11.17; variance = 31.81

Exercise 7.2 page 120

1. **a)** 63.00

b) reading across: 66.32, 63.24, 60.74, 67.62, 60.14, 59.96

c) 64.78, 64.18, 60.05

2. **a)** 61.12

b) 64.5, 63.2, 59.26, 59.7, 59.6, 60.44

c) 63.85, 59.48, 60.02

Q3 is to explore the behavior of repeated samples of different sizes, but there are no solutions here.

Exercise 7.3 page 124

1. **a)** $\overline{x} = 28.9$ $s^2 = 26.4$

b) $\overline{x} = 14.1$ $s^2 = 15.7$

c) $\overline{x} = 69.4$ $s^2 = 11.1$

d) $\overline{x} = 84.8$ $s^2 = 39.2$

e) $\overline{x} = 40.4$ $s^2 = 266$

2. a) $\Sigma f = 51$; $\Sigma xf = 735$; $\Sigma x^2 f = 10699$
$\bar{x} = 14.4$; $s^2 = 2.13$

b) $\Sigma f = 56$; $\Sigma xf = 1540$; $\Sigma x^2 f = 42492$
$\bar{x} = 27.5$; $s^2 = 2.58$

3. a) $\Sigma f = 51$; $\Sigma mf = 732.5$; $\Sigma m^2 f = 12368.75$
$\bar{x} = 14.4$; $s^2 = 36.96$

b) $\Sigma f = 541$; $\Sigma mf = 12995$; $\Sigma m^2 f = 394562.5$
$\bar{x} = 24.0$; $s^2 = 152.6$

4. a) $n = 50$; $\Sigma x = 423$; $\Sigma x^2 = 4956$ $\Rightarrow \bar{x} = 8.46$;
$s^2 = 28.1$

b) $n = 23$; $\Sigma x = 835$; $\Sigma x^2 = 37825$ $\Rightarrow \bar{x} = 36.3$;
$s^2 = 341.4$

c) $n = 68$; $\Sigma x = 695$; $\Sigma x^2 = 12457$ $\Rightarrow \bar{x} = 10.2$;
$s^2 = 79.9$

d) $n = 8$; $\Sigma x = 657$; $\Sigma x^2 = 75688$ $\Rightarrow \bar{x} = 82.1$;
$s^2 = 3104.6$

e) $n = 97$; $\Sigma x = -25$; $\Sigma x^2 = 154$ $\Rightarrow \bar{x} = -0.258$;
$s^2 = 1.54$

5. i) $n = 15$; $\Sigma x = 387$; $\Sigma x^2 = 13221$ $\Rightarrow \bar{x} = 25.8$;
Var$(X) = 215.76$

ii) $s^2 = 231.2$

6. $\bar{y} = 24.0$; $s_y^2 = 5.4$

Exercise 7.4 page 128

1. $(723.4, 760.6)$
2. i) $(602.0, 611.3)$ ii) Narrower
3. $(33.8, 42.2)$
4. 95% CI = $(204.5, 230.5)$ and 98% CI = $(202.1, 232.9)$
5. i) $\bar{x}_{25} = 45.5$ ii) $\sigma = 5.9$ (1 d.p.)

Exercise 7.5 page 131

1. a) $n = 50$; $\Sigma t = 357$; $\Sigma t^2 = 12721$ $\bar{t} = 7.14$;
$w^2 = 207.6$
i) 95% CI is $(3.15, 11.1)$
ii) 90% CI is $(3.79, 10.5)$

b) $n = 37$; $\Sigma x = 1635$; $\Sigma x^2 = 75325$ $\bar{x} = 44.2$;
$w^2 = 85.4$
i) 95% CI is $(41.2, 47.2)$
ii) 90% CI is $(41.7, 46.7)$

c) $n = 62$; $\Sigma v = 651$; $\Sigma v^2 = 12712$ $\bar{v} = 10.5$;
$w^2 = 96.3$
i) 95% CI is $(8.06, 12.9)$
ii) 90% CI is $(8.45, 12.6)$

d) $n = 71$; $\Sigma w = 617$; $\Sigma w^2 = 27688$ $\bar{w} = 8.69$;
$w^2 = 318.9$
i) 95% CI is $(4.54, 12.84)$
ii) 90% CI is $(5.20, 12.18)$

e) $n = 97$; $\Sigma p = -32.4$; $\Sigma p^2 = 174$ $\bar{p} = -0.334$;
$w^2 = 1.670$
i) 95% CI is $(-0.593, -0.075)$
ii) 90% CI is $(-0.552, -0.116)$

2. Not really – a little bit of uncertainly is removed because you do not need to rely on the central limit theorem as an approximating distribution for the sample mean, but you produce the same CI in both cases.

3. $(323.9, 325.3)$ 4. $(27.1, 29.5)$
5. $(502.34, 504.06)$ 6. $(101.1, 103.1)$

Exercise 7.6 page 133

1. a) i) $(0.226, 0.391)$ ii) $(0.239, 0.378)$
b) i) $(0.067, 0.215)$ ii) $(0.079, 0.203)$
c) i) $(0.281, 0.525)$ ii) $(0.301, 0.506)$
d) i) $(0.202, 0.417)$ ii) $(0.220, 0.400)$
e) i) $(0.074, 0.214)$ ii) $(0.086, 0.203)$

2. $(0.056, 0.184)$ 3. $(0.188, 0.387)$

4. i) $(0.429, 0.524)$
ii) that the cars observed form a random sample of all cars on the road (and that the data is accurate – not necessarily easy data to collect)
iii) $\alpha = 99$

5. i) $(0.386, 0.510)$
ii) Need a sample of at least 1521 to have the CI this small.

6. a) i) $n = 242$ ii) $\alpha = 95$

Summary exercise 7 page 134

1. $\Sigma x = 531$; $\Sigma x^2 = 23469$; $n = 13$
$\bar{x} = 40.8$ $s^2 = 148.3$

2. $\Sigma x = 4293$; $\Sigma x^2 = 106979$; $\Sigma f = 173$
$\bar{x} = 24.8$ $s^2 = 2.61$

3. $\Sigma m = 2990$; $\Sigma m^2 = 50525$; $\Sigma f = 212$
$\bar{x} = 14.1$ $s^2 = 39.6$

4. a) $\bar{x} = 3.92$ $s^2 = 31.3$
b) $\bar{x} = 36.3$ $s^2 = 1632$

5. $(73.5, 84.9)$

6. i) $\bar{x}_{35} = 40.8$ ii) $\sigma = 12.4$

7. a) i) 95% CI is $(-2.19, 1.15)$

   ii) 90% CI is $(-1.92, 0.88)$

   b) i) 95% CI is $(10.2, 13.0)$

   ii) 90% CI is $(10.4, 12.8)$

8. $(11.7, 12.9)$ **9.** $(0.037, 0.177)$

10. $(0.158, 0.309)$

11. i) $n = 123$ **ii)** $\alpha = 98.4\%$

12. i) $\bar{d} = 3.9856; \ s^2 = 94.5$

   ii) 96% CI is $(3.979, 3.992)$

   iii) by definition of CIs you expect on average that 96% of the 96% CIs will contain μ, so on average there will be 8 out of 200 which do not.

13. i) $(0.384, 0.553)$

   ii) the people able to be in a shopping centre on a Wednesday afternoon do nto mirror the population you are trying to make inferences about.

14. $(744.4, 751.6)$

15. a) i) Too time consuming; not all pop accessible

   ii) Testing involves destruction

   b) i) 19.7; 0.160 (3 s.f.)

   ii) 0.281   **iii)** Yes

16. $(0.580, 0.630)$ **17.** $n = 35$

18. Smallest n is 20

19. i) 1050; 2304   **ii)** $(1030, 1070)$ **iii)** $n = 246$

20. a) $(39.6, 42.8)$ (3 s.f.)   **b)** $\alpha = 87.5\%$

8 Hypothesis testing for discrete distributions

Skills check page 138

1. $0.056 < 0.1$

2. $x = 5$

Exercise 8.1 page 143

1. $H_0: \ p = 0.23$ *vs* $H_1: \ p > 0.23$

2. Accept H_0 and conclude that there is not sufficient evidence to suggest that the proportion has increased.

3. $H_0: \ \lambda = 0.45$ *vs* $H_1: \ \lambda \neq 0.45$

4. Reject H_0 and conclude that there is evidence at the 5% level of significance to suggest that the average rate of accidents at the weekend is different to weekdays.

Exercise 8.2 page 148

The first few solutions here have extra details included about why the critical region stops where it does at either end of the distribution.

1. two tailed test, critical region includes 0, 1, 9 and 10 and the exact size of the test is 2.1% (including 2 and 8 would increase it to 11%).

2. one tailed test, critical region includes 0 and 1 and the exact size of the test is 1.1% (including 2 would increase it to 5.5%).

3. two tailed test, critical region includes 0 and 10 and the exact size of the test is 0.2% (including 1 and 9 would increase it to 2.1%).

4. two tailed test, critical region includes 0, 8, 9, 10, 11 and 12, and the exact size of the test is 2.3% (including 1 would increase it the bottom tail to 8.5% and including 7 would increase the top tail to 3.9%).

5. one tailed test, critical region includes just 0 and the exact size of the test is 1.4% (including 1 would increase it the size to 8.5%).

6. This example highlights one of the potential problems in testing with small samples – this should be a two tailed test, but to reject at the top end (including only 15 in the critical region) gives a minimum of 3.5% in the top tail. At the bottom end the critical region includes 0–8, with a total probability of 1.8%.

7. one tailed test, critical region includes just 0 and the exact size of the test is 3.5%.

8. would want a one tailed test, but smallest critical region, containing just 0, would have probability of 13.5% so no test is possible.

9. again the two tailed test is impossible to reject at the bottom end, but at the top having the critical region of 6 or more has probability 1.7% (including 5 makes it 5.3%)

10. one tailed test, critical region includes 0 to 3 and the exact size of the test is 3.0% (including 4 would increase it the size to 7.4%).

11. one tailed test, critical region includes 15 and higher, and the exact size of the test is 2.7%.

12. The 2 tailed 10% test will simply include the two 5% one – tailed critical regions and the exact size will be the sum of the two sizes there: critical region is 0 – 3 or 15 and higher, and the exact size is 5.8% (or 5.7% using rounded figures)

Exercise 8.3 page 152

1. i) 16.7%

   ii) 16.7% (the size of the test is the probability of a Type I error)

   iii) Type II error is when null is incorrectly accepted – probability of 2 or more is 45.1%

2. i) 0, 1   ii) 1.7%   iii) 1.7%

   iv) Type II error is when null is incorrectly accepted – probability of 2 or more is 80.0%

3. i) 3.5%   ii) 3.5%

   iii) Type II error is when null is incorrectly accepted – probability of 3 or more is 32.3%

4. a) i) 3 or more   ii) 4.7%

   b) i) reject H_0 and conclude there is sufficient evidence at the 5% level to suggest that the mean rate is greater than 0.8.

   ii) Type I – since the test rejects H_0 the only error possible is if H_0 were true.

5. i) $P(Po(3) = 0) = 0.0498$; $P(Po(3) = 0,1) = 0.199$, so the critical region for a 5% test only contains 0.

   ii) $P(Po(0.3) = 0) = 0.741$; $P(Po(0.3) > 0) = 0.259$, so if $\lambda = 0.3$ the probability of a Type II error (accepting the null hypothesis incorrectly) is 25.9%.

6. a) i) $P(X \geq 16) = 5.1\%$; $P(X \geq 17) = 1.6\%$ so the critical region for a 5% test is $X \geq 17$.

   ii) 1.6%

   b) 75% of 20 is 15, so the test would not reject H_0 meaning that a Type I error is the only type that could have occurred.

Exercise 8.4 page 155

1. If she was guessing then the number of correct answers, X, she gets on the test can be modelled by the $B(15,0.2)$ distribution.

   The test is $H_0: p = 0.2$ vs $H_1: p > 0.2$, and the critical region is 7 or more $P\{B(15, 0.2) \geq 6\} = 6.1\%$; $P\{B(15, 0.2) \geq 7\} = 1.8\%$ so there is not sufficient evidence to suggest that she has done better than by guessing.

2. There is nothing to suggest that more heads were expected before knowing that 10 out of 12 tosses were heads, so you need to look at the two tailed test – i.e. reject null hypothesis if 0, 1, 2, 10, 11 or 12 heads are seen since $P\{B(12, 0.5) = 0, 1, 2, 10, 11, 12\} = 3.9\%$;

   i) Since 10 heads were observed, reject H_0 and conclude that there is evidence to suggest the coin is biased.

ii) Type I is the only one possible since H_0 was rejected by the test.

3. If X is the number of flights arriving more than 5 minutes the published arrival time, then $X \sim B(30, 0.04)$. $P(X \geq 2) = 11.7\%$; $P(X \geq 3) = 3.1\%$ so the (one tailed) test, at the 10% level of significance, is 'Accept H_0 if no more than 3 flights are more than 5 minutes late'

   Since 3 flights in the sample arrived more than 5 minutes after the scheduled arrival time, H_0 is accepted and you can conclude there is insufficient evidence to suggest that the airline has overstated the proportion of flights arriving within 5 minutes of the published arrival time.

4. Since the proportion arriving on time in the sample is 91.4% which is greater than the 90% level that the company claims, it cannot be evidence that they overstate their performance.

5. i) the critical region for a 10% test will always include all the critical region for the 5% test (and except for very small sample tests will be larger) – so the 10% test will be more likely to conclude the financial advisor has overstated their performance.

   ii) the test is of $H_0: p = 0.1$ vs $H_1: p > 0.1$; $P\{B(20, 0.1) \geq 4\} = 13.3\%$; $P\{B(20, 0.1) \geq 5\} = 4.3\%$ so the critical region is 5 or more. Since 4 in the observed sample were not satisfied you accept H_0 and conclude there is not sufficient evidence to suggest that the financial advisor has overstated their performance.

   iii) since H_0 was accepted, a Type II error could have been made.

6. $H_0: p = 0.25$ vs $H_1: p < 0.25$

   i) $P\{B(24, 0.25) \leq 2\} = 4.0\%$; so the significance level of the test is 4.0%

   ii) 4.0%

   iii) $P\{B(24, 0.1) \leq 2\} = 56.4\%$; so the probability of accepting H_0 incorrectly when $p = 0.1$ is 56.4%.

Exercise 8.5 page 157

1. $H_0: \lambda = 9.5$ vs $H_1: \lambda < 9.5$

   i) $X \sim Po(9.5)$. $P(X \leq 4) = 4.0\%$; $P(X \leq 5) = 8.9\%$, so the critical region is 0–4.

   ii) 4.0%

   iii) This value lies in the critical region, so the null hypothesis is rejected and you conclude at the 4% (or 5%) level that there is evidence to suggest that the parameter is lower than 9.5.

2. i) On average in a half hour period there would be 10 hits, and the conditions for a Poisson (independence and randomness) are likely to be satisfied, so $X \sim \text{Po}(10)$ is the appropriate distribution.

ii) $P(X \leq 4) = e^{-10}\left(1 + 10 + \frac{10^2}{2!} + \frac{10^3}{3!} + \frac{10^4}{4!}\right) = 0.029$

iii) If it is a site that is likely to be of more interest to people in a area of the world with similar time zones (e.g. of interest mainly in Asia, or in Europe, or in the Americas), then there might be more activity on the site during the day in that region than at nighttime.

iv) $P(X \leq 4) = 0.029$, so 4 lies in a one tailed test critical region but not in the two tailed 5% test. The reasoning in part **iii)** suggests that a one tailed test for a decrease in the rate during the night is appropriate, but without it, the two tail test would be the one to use.

3. a) The conditions for a Poisson (independence and randomness) are likely to be satisfied, so $X \sim \text{Po}(3)$ is the appropriate distribution.

b) $P(X \geq 5) = 1 - e^{-3}\left(1 + 3 + \frac{3^2}{2!} + \frac{3^3}{3!} + \frac{3^4}{4!}\right) = 0.185$

c) If Y is the number of months in a year in which 5 or more accidents occur then $Y \sim \text{B}(12, 0.185)$, $P(Y \geq 1) = 1 - 0.815^{12} = 0.914$. So the probability the crossing will be installed is 0.914.

d) using a 5% significance level (one tailed test for a decrease) gives critical regions **i)** for 1 month – 0 accidents (5.0% level) and **ii)** for 3 months up to 3 accidents (2.1% test)

e) i) the 1 month test period is quicker, but **ii)** the 3 month test period is likely to be better in detecting a shift in the average rate (a lower Type II error for the same significance level is the reward for using a larger sample in a test).

4. i) $X \sim \text{Po}(2.5)$: $P(X \geq 5) = 10.9\%$; $P(X \geq 6) = 4.2\%$, so the critical region for a 5% test is 6 or more.

ii) 4.2%

iii) This does not lie in the critical regions, so accept H_0 and conclude that there is not sufficient evidence at the 5% level to suggest that the parameter is greater than 2.5.

5. The conditions for a Poisson (independence and randomness) are likely to be satisfied, so $X \sim \text{Po}(2.5)$ is the appropriate distribution.

ii) $P(X \leq 4) = e^{-2.5}\left(1 + 2.5 + \frac{2.5^2}{2!} + \frac{2.5^3}{3!} + \frac{2.5^4}{4!}\right) = 0.109$

iii) The behavior just before it closes might be affected by people deciding not to come in because they would not have time to do all they wanted – or making a special effort to get there because they have been working all day and it is the only chance they have – not possible to say if it will be but enough differences in conditions to suggest that the average rate might be different.

iv) the reasoning in **iii)** says that a two tailed test is appropriate: $P(X \geq 5) = 4.2\%$ (>2.5%) so accept H_0 and conclude that there is not sufficient evidence at the 5% level to suggest that the average rate is different in the last half hour.

6. a) The conditions for a Poisson (independence and randomness) are likely to be satisfied, so $X \sim \text{Po}(2.2)$ is the appropriate distribution.

b) $P(X \geq 5) = 1 - e^{-2.2}\left(1 + 2.2 + \frac{2.2^2}{2!} + \frac{2.2^3}{3!} + \frac{2.2^4}{4!}\right)$

$= 0.0725$

c) If Y is the number of months in a year in which 5 or more accidents occur then $Y \sim \text{B}(12, 0.0725)$, $P(Y \geq 1) = 1 - 0.9275^{12} = 0.595$. So the probability the crossing will be installed is 0.595.

d) i) using a 5% significance level (one tailed test for a decrease) it is not possible to construct a test using only one month since the probability of no accidents in a month is 11.1%

ii) for 3 months up to 2 accidents (4.0% test)

e) i) the 1 month test period would be quicker, but is not possible. **ii)** a longer test period is always better in detecting a shift in the average rate and in this instance there is no choice to be made because a one month period is not long enough even to get a 5% test.

Summary exercise 8 page 159

1. i) one tailed 5% test has critical region of just 0 – exact significance level is 4.3%

ii) two tailed 105% test has critical region of 0, 1 or 2 (4.3% at the bottom tail) and 12 or higher (3.4% at the top tail) – so the exact significance level of the test is 7.7%.

2. i) one tailed 2% test has critical region of just 0.

ii) exact significance level is 1.1%

iii) 1.1%

iv) Type II error is when the null hypothesis is incorrectly accepted – here the probability of getting 0 (the only value in the critical region) is 13.5% so P(Type II error) is 86.5%

3. i) the test is of H_0: $p = 0.95$ *vs* H_1: $p < 0.95$; $P\{B(20, 0.95) \le 16\} = 0.016 < 5\%$ so you reject H_0 and conclude there is evidence at the 5% level to suggest that the council has overstated the level of satisfaction.

ii) since H_0 was rejected, a Type I error could have been made.

4. i) the test is of H_0: $p = 0.1$ *vs* H_1: $p > 0.1$, and $P\{B(20, 0.1) \le 4\} = 0.957$ so the critical region for the test is 'at least 5 out of the 20 approve', so the test will accept that the proportion has increased.

ii) This still only represents $\frac{1}{4}$ of the residents asked about the revised plans so it is hardly a resounding endorsement – just because it has increased the approval rate does not mean that enough people approve it for it to be a good idea to proceed.

5. i) $P\left\{B\left(20, \frac{1}{6}\right) \le 1\right\} = 0.130 > 5\%$ so accept H_0 and conclude that there is not sufficient evidence to suggest that ones appear less often than on a fair die.

ii) the critical region for the test is just 0, and the probability that this is seen when p does equal $\frac{1}{6}$ is 0.026, so the probability of a Type I error is 2.6%.

iii) a type II error is to accept that ones do not appear less often than in a fair die when they do appear less often.

6. i) $P\{Po(1.2) \ge 4\} = 0.034$; $P\{Po(1.2) \ge 3\} = 0.121$ so the rejection region (=critical region) is 4 or higher.

ii) Type II error is to accept the null hypothesis incorrectly – accepting H_0 here is if no more than 3 hurricanes are observed: $P\{Po(2.5) \le 3\} = 0.758$, so there is a 75.8% chance of making a Type II error if the mean number of hurricanes has increased to 2.5 per year.

7. i) 0.0342   **ii)** Accept H_0 **iii)** 0.915

8. i) Accept claim   **ii)** 0.0292

9. i) 0.0202   **ii)** 0.972   **iii)** 0.0311

10. i) CR is $X \le 1$ **ii)** 0.0134

iii) No evidence the mean number of injuries has decreased.

iv) 0.128 (3 s.f.)

9 Hypothesis testing using the Normal distribution

Skills check page 162

1. $x = 63.8$

Exercise 9.1 page 166

Solutions for questions 1 and 2 are shown by both methods below:

Method 1:

1. i) reject H_0.   **ii)** accept H_0.   **iii)** reject H_0.

2. i) reject H_0.   **ii)** accept H_0.   **iii)** reject H_0.

Method 2:

1. i) reject H_0 **ii)** accept H_0 **iii)** reject H_0

2. i) reject H_0 **ii)** accept H_0 **iii)** reject H_0

3. i) comparing the first and second tests in each question: the population has a larger variance in the second one and the acceptance region includes more values – it is harder to be sure that an observed difference would not occur because of the randomness in the sampling process – and the same observed mean results in **ii)** being rejected where **i)** was accepted.

ii) comparing the second and third tests in each question: the sample size in the third one is larger and the acceptance region shrinks – with more data it is easier to be sure that the observed difference would not occur because of the randomness in the sampling process – and the test goes back to accepting the null hypothesis.

4. accept H_0

5. reject H_0

6. H_0: $\mu = 85.3$ *vs* H_1: $\mu \ne 85.3$ are the hypotheses for the test, with $\sigma = 3.2$

Plan A:

i) $n = 5$, so 5% test is Accept H_0 if
$$\bar{x} \in \left(85.3 - 1.96 \times \frac{3.2}{\sqrt{5}}, \ 85.3 + 1.96 \times \frac{3.2}{\sqrt{5}}\right)$$
$$= (82.5, 88.1)$$

ii) $P\left\{N\left(87.3, \frac{3.2^2}{5}\right) \in (82.5, 88.1)\right\} = 0.712$

Note that using a downward shift in the mean of 2 hours would give the same answer because of the symmetry of the situation (the top / bottom tail probabilities would be reversed).

Plan B:

i) $n = 30$, so 5% test is Accept H_0 if
$$\bar{x} \in \left(85.3 - 1.96 \times \frac{3.2}{\sqrt{30}}, \ 85.3 + 1.96 \times \frac{3.2}{\sqrt{30}}\right)$$
$$= (84.2, 86.4)$$

ii) $P\left\{N\left(87.3, \frac{3.2^2}{30}\right) \in (84.2, 86.4)\right\}$

$$= P\left\{\frac{84.2 - 87.3}{\frac{3.2}{\sqrt{30}}} = -5.31 < z < \frac{86.4 - 87.3}{\frac{3.2}{\sqrt{30}}} = -1.540\right\}$$

$$= \Phi(-1.540) - \Phi(-5.36) = 0.062 - 0.000 = 0.062$$

Note – this illustrates why taking large samples is so much more effective in discriminating when there has been an important shift in the mean – the probability of accepting the null hypothesis incorrectly has been reduced from 71% to just 6%.

iii) **Note:** this is not something you would be asked to do in an examination.

A sample size 49 is the smallest which meets the MD's request.

Exercise 9.2 page 169

1. i) Accept H_0 and conclude there is not sufficient evidence at the 5% level to suggest that the mean is not 45.

   ii) Reject H_0 and conclude there is evidence at the 5% level to suggest that the mean is less than 57.5.

   iii) Accept H_0 and conclude there is not sufficient evidence at the 5% level to suggest that the mean has increased from −0.5.

2. i) Accept H_0 conclude there is not sufficient evidence at the 5% level to suggest that the proportion is not 0.4.

   ii) Accept H_0 and conclude there is not sufficient evidence at the 1% level to suggest that the parameter is less than 27.1

   iii) There is not sufficient evidence at the 5% level to suggest that the proportion has increased from 0.8

Note – if you calculated the observed proportion was <0.8, you could have drawn this conclusion without doing the full calculations.

3. There is not sufficient evidence at the 5% level to suggest that the proportion s greater than 0.2.

4. There is not sufficient evidence at the 1% level to suggest that the average rate of accidents has decreased.

5. i) $z = \frac{\bar{x} - \mu}{\frac{s}{\sqrt{n}}} \Rightarrow 1.867 = \frac{75.2 - 74.0}{\frac{5.3}{\sqrt{n}}}$

   $\Rightarrow \sqrt{n} = \frac{1.867 \times 5.3}{75.2 - 74.0} = 8.246 \Rightarrow n = 68$

   ii) for a 5% one tailed test the cv is 1.645, so reject H_0 and conclude that there is evidence at the 5% level to suggest that the mean height has increased.

6 a) i) 0.864   ii) 0.824

   b) Around holiday periods and birthdays, there is likely to be more letters than usual.

   c) Accept H_0 and conclude there is not sufficient evidence to suggest the mean has increased.

Exercise 9.3 page 171

1. a) Accept H_0. b) Accept H_0. c) Accept H_0.

2. i) (599.97, 601.63)

   ii) Does not support the claim that the mean contents are 602 ml.

3. i) (247.6, 249.4)

   ii) Does not support the claim that the mean contents are 250 g.

Summary exercise 9 page 171

1. i) while the decrease might get them into legal difficulties, an increase in the mean contents will cost them money.

   ii) There is evidence at the 5% level to suggest that the mean has changed.

2. Reject H_0 and conclude that there is evidence at the 5% level to suggest that the pulse rates of trained athletes is lower than that of healthy young adults.

3. i) to take measurements of a complete population can be very expensive and take a long time – a sample is cheaper and offers quicker access to a reasonable level of information.

ii) Reject the claim at the 5% level of significance.

iii) What the spokesperson says may be true – and you can construct scenarios which give plausible explanations why this may not be reflected in the customers' experiences – for instance if they reduced very slightly the 40% most expensive holidays that very few people actually bought last year then they could make the claim accurately even if it presents a misleading picture to the customer.

4. There is evidence at the 5% level to suggest that the average rate of accidents has decreased.

5. There is not sufficient evidence at the 1% level to suggest that the average height has increased.

6. i) $H_0: \mu = 15$ or $p = 0.25$; $H_1: \mu > 15$ or $p > 0.25$

ii) Justified

7. i) $H_0: \mu = 21.2$; $H_1: \mu = 21.2$. Not the same

ii) Not the same sentence length or author; P(Type I error) = 5%

8. Accept teacher's estimate

9. i) $H_0: \mu = 46$; $H_1: \mu = 46$

ii) No significant difference in times

Review exercise C page 174

1. (98.8, 99.6)

2. i) (29.4, 32.6) **ii)** Accept claim at 2% level

3. i) $n = 600$ **ii)** $\alpha = 97.4\%$

4. i) $\mu = 227.1$; $\sigma^2 = 265$

ii) $n = 78$

5. i) There are 20% red chocolate beans but George says there are fewer

ii) 0.167

6. i) Not random

ii) No significant improvement in survival rate

iii) Saying no improvement when there is

iv) 0.950

7. i) $n = 250$

ii) $z = 1.646$; 90% confidence interval

8. i) A sample where every element has an equal chance of being chosen.

ii) (0.321, 0.422) **iii)** $n = 2241$

9. i) 4.125 **ii)** 330 to 332

iii) No; 333 is not within CI

10. ii) 0.0172 (3 s.f.)

11. i) Excludes groups of people, e.g. children, people without phones etc.

ii) 0.119 to 0.261 (3 s.f.) **iii)** $x = 93$ (3 s.f.)

12. Not enough evidence of change.

13. i) 0.0480 **ii)** 0.0480 **iii)** 0.601

14. i) 6.53; 2.87

ii) Reduction in cars exceeding speed limit.

iii) Stating there is a reduction in the number of speeding cars when there isn't

15. i) 0.0234 **ii)** 0.160

16. i) 1.2637

ii) No evidence that mean has changed

iii) No, because H_0 not rejected

iv) Stating mean not changed when it has; $-1.96 <$ test stat < 1.96

17. i) Not enough evidence to support claim.

ii) Stating no extra weight loss when there is.

18. i) 150 **ii)** Evidence that μ has increased.

Index

Page numbers in *italics* refer to Summary Questions.